Universität Stuttgart
Institut für Energieübertragung und Hochspannungstechnik, Band 34

Estimation of Flexibility Potentials
in Active Distribution Networks

Estimation of Flexibility Potentials in Active Distribution Networks

Von der Fakultät

Informatik, Elektrotechnik und Informationstechnik

der Universität Stuttgart

zur Erlangung der Würde eines Doktor-Ingenieurs (Dr.-Ing.)
genehmigte Abhandlung

vorgelegt von

Daniel Alfonso Contreras Schneider

aus Santiago, Chile

Hauptberichter:	Prof. Dr.-Ing. habil. K. Rudion
Mitberichter:	Prof. Dr.-Ing. J. Myrzik
Tag der Prüfung:	04.10.2021

Institut für Energieübertragung und Hochspannungstechnik

der Universität Stuttgart

2021

Bibliografische Information der Deutschen Nationalbibliothek:

Die Deutsche Nationalbibliothek verzeichnet diese Publikation in der Deutschen Nationalbibliografie, detaillierte bibliografische Daten sind im Internet über http://dnb.dnb.de abrufbar.

Universität Stuttgart
Institut für Energieübertragung und Hochspannungstechnik, Band 34

D 93 (Dissertation Universität Stuttgart)

Estimation of Flexibility Potentials in Active Distribution Networks

Herstellung und Verlag: BoD – Books on Demand, Norderstedt
ISBN: 978-3-75439-704-6

Preface

The most exciting phrase to hear in science, the one that heralds new discoveries, is not "Eureka!" but "That's funny…", stated Isaac Asimov, my favorite author. This very well relates to the thesis you are about to read, which is not the product of momentary genius, but rather a conjunction of gradual improvements made upon existing models. What Asimov perhaps meant is that current models and ideas should always be questioned, in order to find a way to improve them. It was in 2017, during a dialog with my first student, Bertin Wagner, that I realized that the aggregation algorithm we were trying to replicate could be significantly improved by incorporating ideas obtained from my colleagues. The work of Bertin allowed me to demonstrate these improvements to the model, and I thank him for that. The next pages will tell you what happened from that "that's funny" moment on.

Nonetheless, none of this would have been possible without the help of many wonderful persons all over the world. Beginning with Prof. Dr. Krzysztof Rudion and Prof. Dr. Stefan Tenbohlen, to whom I am grateful for the opportunity of pursuing the PhD at the IEH, and for all the support during my time there. The contribution of Prof. Dr. Johanna Myrzik to this work is greatly appreciated, most of which came during a casual hallway chat the first time we met at a conference.

A large number of great persons crossed my path during my time working on the research projects CALLIA, FELSeN and flexQgrid, as well during the preparation of several other project proposals. They were of great help for the development of this work.

For the talented Kathrin Walz and Dr. Pascal Wiest, my amazing office partners, I have nothing but gratitude and admiration. The same is extended to Heiner Früh, who proof-read my thesis and encouraged me to submit it without hesitation. The IEH is full of superb individuals, all of whom deserve my appreciation, including all my coworkers, the people at the workshop and all the administrative staff. Special thanks go to Dr. Ulrich Schärli for all his patience and help since 2009.

I would like to thank my parents Heloisa and Eduardo and my sister Ana, wo were always supporting me from halfway around the world. Finally, this work is dedicated to my late grandparents Valmira, Alfons and Leonardo, as well as my grandmother Ulrika. Their hard work, perseverance and honesty will always be an inspiration for me. This adventure would not have taken place without them, including the ones to come…

Abstract

Replacing conventional power plants by distributed energy resources (DER) in the MV and LV grids poses great new challenges for the planning and operation of distribution grids. Controlling a handful of conventional power plants demands significantly less resources than operating numerous decentralized plants, especially when in involves the supply of ancillary services to maintain the grid stability. Nevertheless, most ancillary services are required at the transmission system level, meaning that vertical supply of flexibility becomes necessary, requiring new methods to quantify how much flexibility can be provided from distribution (DSO) to transmission system operators (TSO). The feasible operation region (FOR) allows capturing the aggregated flexibility potential of DER within a distribution grid, while respecting the technical restrictions of both plants and grid.

This thesis proposes a novel approach to compute the FOR, the Linear Flexibility Aggregation (LFA) method, based on the solution of linear OPF. With the objective of reducing the computation time, without compromising the accuracy of the assessed FOR.

The LFA algorithm is comprehensively evaluated throughout this thesis, focusing on the accuracy of speed of the approach. The analysis quantifies the impact of the linear OPF model in the FOR computation, as well as it identifies all relevant parameters that can have an impact in the computation time.

It is shown that the proposed method provides a considerable reduction in processing time compared to similar methods, e.g. Monte-Carlo simulations or non-linear OPF-based methods. The linearization of the power flow equations has an impact in the accuracy of the solution, however, the trade-off with the reduction of the computational time is acceptable.

The dissertation closes with the suggestion of three use cases for the LFA method. Firstly, it is described how a fast computation of the FOR could be used to study the long-term provision of flexibility in distribution grids. Secondly, the usage of the LFA for the vertical aggregation of flexibility over different voltage levels is shown. Finally, the inclusion of the FOR concept in a congestion management approach, including redispatch at the distribution grid level is demonstrated. The proposal and analysis of these use cases applied to large distribution grids would not have been possible without a fast and reliable computation algorithm, like the LFA.

Overall, the coordination between grid operators can benefit significantly from the fast computation of the FOR, allowing its inclusion not only in planning processes, but also in everyday operation processes.

Kurzfassung

Die Ersetzung von konventionellen Kraftwerken durch eine Vielzahl verteilter Energieerzeugungsanlagen in den Mittel- und Niederspannungsnetzen stellt große neue Herausforderungen für Planungs- und Betriebsprozesse von Verteilnetzen dar. Einige wenige konventionelle Kraftwerke zu steuern erfordert deutlich weniger Ressourcen, als der Betrieb einer Vielzahl dezentraler Anlagen, um deren Flexibilität für Netzdienstleistungen zu nutzen. Die meisten Netzdienstleistungen werden jedoch auf der Übertragungsnetzebene benötigt, so dass ein vertikaler Stromtransport notwendig ist, was neue Methoden erfordert, um zu quantifizieren, wie viel Flexibilität auf der Verteilnetzebene bereitgestellt werden kann.

Die Feasible Operation Region (FOR) ermöglicht die Erfassung des aggregierten Flexibilitätspotenzials von DEA innerhalb von Verteilnetzen, wobei die technischen Restriktionen sowohl der Anlagen, als auch des Netzes berücksichtigt werden.

Diese Arbeit schlägt eine neuartige Methode zur Berechnung der FOR vor, die LFA-Methode (Linear Flexibility Aggregation), die auf der Lösung mehrerer völlig linearer OPF basiert. Dieser Ansatz bietet im Vergleich zu ähnlichen Methoden, z.B. Monte-Carlo-Simulationen oder nicht-linearen OPF-basierten Methoden, eine erhebliche Reduktion der Rechenzeit. Die Linearisierung der Leistungsflussgleichungen wirkt sich zwar auf die Genauigkeit der Lösung aus, der Kompromiss mit der Reduzierung der Rechenzeit ist jedoch bedeutend.

In dieser Arbeit wird die Anwendung des LFA-Algorithmus aus drei verschiedenen Perspektiven analysiert. Zunächst wird die Qualität des Ansatzes sorgfältig untersucht, um die Auswirkungen des linearen Modells bei der FOR-Berechnung zu quantifizieren und alle relevanten Parameter zu identifizieren, die die Berechnungszeit beeinflussen. Dann wird die Anwendung der Methode zur Analyse der langfristigen Bereitstellung von Flexibilität eines Verteilnetzes demonstriert. Schließlich wird die Einbeziehung des FOR-Konzepts in einen Engpassmanagement-Ansatz, einschließlich Redispatch auf der Verteilnetzebene, demonstriert, wo die Vorteile des schnellen Algorithmus zum Tragen kommen.

Allgemein kann die Koordination zwischen Netzbetreibern erheblich von der schnelleren Berechnung des FOR profitieren, so dass es nicht nur in Netzplanungs-, sondern auch in alltägliche Betriebsprozesse einbezogen werden kann.

Table of Contents

Abbreviations

Abbreviation	Meaning
AC	Alternating current
ADN	Active distribution network
BESS	Battery energy storage system
CHP	Combined heat-power
CIG	Converter-interfaced generation
DC	Direct current
DCOPF	Direct current optimal power flow
DG	Distributed generation
DFIG	Doubly-fed induction generator
DSI	Demand side integration
DSO	Distribution system operator
EENS	Expected energy not served
ESS	Energy storage system
EV	Electrical vehicle
FOR	Feasible operation region
FCWT	Full-converter wind turbine
FPG	Flexibility providing grid
FPU	Flexibility providing unit
FRT	Fault-ride-through
FXOR	Flexible operation region
HVAC	High voltage air conditioning
HVDC	High voltage direct current
HV	High voltage
ICPF	Interval Constrained Power Flow (Algorithm)
IG	Induction generator
IPF	Interconnection power flow
IRRE	Insufficient ramping resource expectation

LFA	Linear Flexibility Aggregation (Algorithm)
LLU	Loss of largest unit
LOLE	Loss of load expectation
LV	Low voltage
MPPT	Maximum power point tracker
MV	Medium voltage
NCP	Network coupling point
NR-PF	Newton-Raphson power flow (algorithm)
OLTC	On-load tap changer
OPF	Optimal power flow
PCC	Point of common connection
PCR	Primary control reserve
PDF	Probability distribution function
PV	Photovoltaic
PWM	Pulse-wide modulation
RES	Renewable energy sources
RS	Random sampling
SCOPF	Security constrained optimal power flow
SG	Synchronous generator
TSO	Transmission system operator
VPP	Virtual power plant
VSC	Voltage source converter
WG	Wind generator

Symbols

Symbol	Meaning
A_{eq}	Matrix with left-hand-side of linear equalities
A_{ineq}	Matrix with left-hand-side of linear inequalities
$area(P)$	Function that computes the area of a polygon
a, b, c	Coefficients of straight-line equation
b_{eq}	Vector with right-hand-side of linear equalities
b_{ineq}	Vector with right-hand-side of linear inequalities
$c_{eq}(x)$	Non-linear equality constraints in optimization problem
$c_{ineq}(x)$	Non-linear inequality constraints in optimization problem
$conv(P)$	Convex hull of polytope
γ, δ	Search direction parameters of LFA objective function
$\underline{E}$	Complex voltage at slack bus
E_{TH}	Thevenin equivalent voltage magnitude
$f(x)$	Objective function of optimization problem
$F(x)$	Multivariate function
F_i	Set of FPU connected to a bus
FOR	Polygon representing the FOR of a grid
$\underline{I}, \underline{i}$	Complex current
I_{th}	Thermal current limit of power line
J	Jacobian matrix of a multivariable function
J^{-1}	Inverse of Jacobian matrix of a multivariable function
L	Straight-line equation describing a segment
$\mathcal{N}(\mu, \sigma^2)$	Normal distribution with mean μ and variance σ^2
$NRPF(x)$	Newton-Raphson Power Flow calculation
p_{fix}, q_{fix}	Fix active and reactive power consumption at bus
p_{FPU}, q_{FPU}	Flexible active and reactive power consumption of FPU
P_G, Q_G	Generator operation point (active and reactive power)
p_i, q_i	Calculated active/reactive power injection at bus
$\tilde{p}_i, \tilde{q}_i$	Injected active/reactive power at bus

P_{ij}, Q_{ij}	Calculated active/reactive branch power flow
$\Delta P_i, \Delta Q_i$	Mismatch between injected and calculated power
$\Delta P_{ij}, \Delta Q_{ij}$	Deviation of active/reactive branch power flow
$P_{PV}(t)$	Injected active power by PV generator in function of time
$P_{PV,max}$	Rated capacity of PV generator
ϑ_i	Bus voltage angle
ϑ_{ij}	Voltage angle difference between two buses
$\Delta\vartheta_{i,k}$	Difference of voltage angle for bus
$S_{ij,max}$	Max. apparent power of branch
$SOC(t)$	SOC of the storage system in function of time
SOC_{max}, SOC_{min}	Max. and min. SOC allowed by the storage system
$similarity(P_1, P_2)$	Index indicating similarity between two polygons
T_f	Number of sides of a polygon
$\underline{t}$	Complex transformation ratio of a transformer
$\underline{U}, \underline{u}$	Complex bus voltage
u_i	Bus voltage magnitude
$\Delta u_{i,k}$	Difference of voltage magnitude for bus
u_{min}, u_{max}	Bus voltage magnitude min. and max. boundaries
v_f	Vertex of polygon (complex power)
w_{ij}	Switching state of branch
x	State variables vector
x_{ub}, x_{lb}	Upper/Lower boundaries of state vector
Δx_k	Correction factor of state variables vector
Y	Admittance matrix
$y_{ij}\angle\theta_{ij}$	Complex branch admittance
$\underline{y}_T$	Complex admittance of a transformer
$\underline{Z}, Z\angle\theta$	Power line complex impedance
$Z_{TH}\angle\Theta_{TH}$	Thevenin equivalent complex impedance

1   Introduction

1.1   Motivation and Background

Flexibility is perhaps one of the most echoed concepts within the electrical power systems community in the last decade. It might be perceived as a concept belonging to modern times, nonetheless, it is one of the key features of power systems since the beginnings. Otherwise, it would not be possible for a power system ever to have become the "largest interconnected system ever built by Men" [1]. Flexibility is an essential attribute of any electrical grid, otherwise it is impossible for them to have a stable operation, especially when composed of thousands, if not millions, of mechanical and electronical components.

Massive conventional power plants have been the main providers of flexibility for many decades, and that has been sufficient to keep the power system running in a secure manner. Nevertheless, recent years have seen renewable energy sources (RES) and energy storage systems (ESS) become more economically viable, demand side integration (DSI) reach household applications, electric vehicles (EV) disrupt markets around the globe, and old-fashioned Ferraris meters replaced with smart meters. This means that power systems are evolving, and consequently all planning and operation strategies involved need to evolve accordingly.

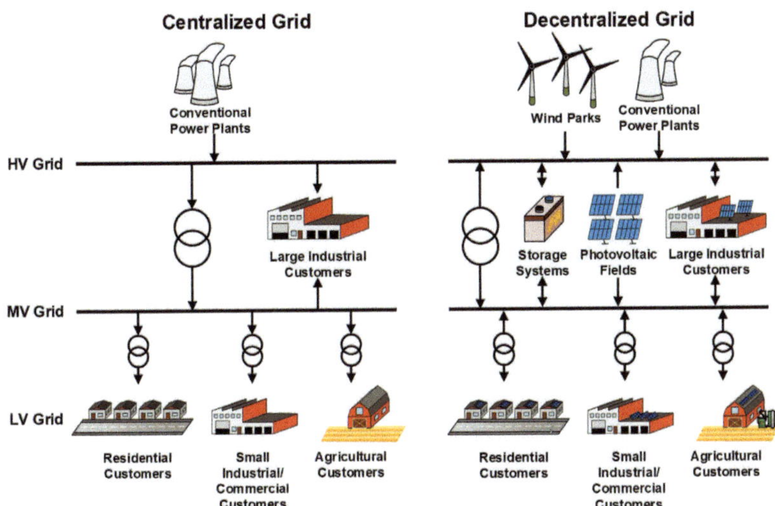

Figure 1-1:   The changing landscape of energy systems. [2]

Modernity brings new challenges along; RES generation is volatile and challenging to predict, new loads are causing the overall power consumption to increase (e.g. electrification of transport and heating), there is an increasing opposition against the construction of new power lines and power plants (including RES), and conventional power plants are being decommissioned following economic and political reasons. These are just a few examples of changes. The traditional top-down power flow from the centralized generation to the customers is changing, as the share of distributed energy resources (DER) at the lower voltage levels is increasing (Figure 1-1). This forces a redefinition of the role of active distribution networks (ADN) [3]. Additional challenges arise when it comes to ensuring a stable operation of the entire power system, as traditional facilitators of ancillary services are diminishing, while new technology and services providers are surfacing. Thus, flexibility requirements for the provision of ancillary services to the grid (e.g. frequency control, congestion management, voltage control) are changing as well.

A noteworthy paradigm change is that the operation of a handful of conventional power plants demands significantly less technological resources than operating a large number of small-scale DER, which can provide flexibility to the grid. These pieces of equipment are defined as flexibility providing units (FPU) [4]. Therefore, the operation and control of power systems is becoming even more complex, especially at the distribution level. Yielding as much flexibility as possible from decentralized FPU requires a great deal of coordination within the organization of grid operators, and among them as well. One extreme case is the German power system, where each of the four transmission system operators (TSO) needs to coordinate with over a hundred distribution grid operators (DSO) within their control zones. This scenario hampers for any TSO to achieve a proper level of coordination with all its underlayered DSO. Additionally, the unbundling of the electrical sector adds a supplementary burden to the communication among grid operators, requiring new methods to strengthen TSO-DSO cooperation. For example, grid operators are in some cases not allowed to share details of their grid models between them, or with other stake holders (e.g. aggregators, electricity markets).

To improve the cooperation between different stake-holders, novel methods need to be developed that enable the communication of the grid state, including the aggregated capability to provide flexibility for ancillary services. These methods should help the grid operators to avoid revealing sensitive grid information. As the operation of power systems is time-critical, it requires fast-acting solutions, therefore, all developed instruments not only need to be effective, but also computationally efficient. This acts as the main motivation of the presented thesis, which focuses on "optimizing the computation of the feasible operation region in distribution grids".

1.2 State-of-the-Art and Objectives

For over a century, the operation of synchronous generators (SG) has been carefully analyzed using different types of methods. Many of these are based on graphical approaches, which became known as "capability charts", describing the set of active and reactive power points in which the machines can operate is safety [5]. In [6], a subtle, yet significant modernization to the concept was proposed and capability charts were suddenly not only suitable for the analysis of electrical machines, but also to represent the feasible operation region (FOR) of a power system [7]. Since then, plenty of research has been dedicated to develop automatized and computationally efficient methods to compute the FOR of a grid, e.g. [8], [9], [10], [11], [12], and [13]. The FOR, however, describes only the technically feasible operation points that could be achieved by DER. The authors of [14] correctly stated that both concepts, feasibility and flexibility, are not to be mistaken. This led to the definition of the flexibility operating region (FXOR). This is relevant for this thesis, which focuses on the flexibility provision of ADN, with flexibility defined as the "modification of generation or consumption in reaction to an external signal" [15]. One important aspect when it comes to the FOR and FXOR concepts, is that they focus not only on active power flexibility, but also on reactive power flexibility. Reactive power is in many studies and applications either neglected or simplified, especially as no specific market mechanism for its application exists. With the decommission of large conventional power plants, the main suppliers of reactive power until now, DER will be compelled to increase their role in the provision of reactive power. This topic needs to be addressed properly, especially when it comes to TSO-DSO cooperation schemes, e.g. [11], [16], [17], [18], [19], and [20].

This aspect motivated linking the FOR concept into TSO-DSO coordination strategies in [21]; where a Monte-Carlo-based computation method was proposed. In order to solve the limitations of random sampling simulations regarding processing time, the advanced Interval Constrained Power Flow (ICPF) algorithm was proposed in [22] and [23]. The ICPF algorithm computes the FOR by means of solving a set of nonlinear mixed-integer optimal power flows (OPF). This idea was not completely new, as the usage of OPF to calculate the FOR was already covered in the early nineties in [8] and [9]. At the time, the limited computational power restricted the possibilities to develop such methods further, as solving a nonlinear OPF is a complex task, which can demand hefty computation power. Additionally, when analyzing larger grids, a non-linear OPF can have convergence issues or even numerical errors. The novelty of the ICPF algorithm was the optimization of the number of OPF that need to be solved to properly assess a FOR. However, the required processing time of that algorithm may still be too large under certain conditions, as can need over 15 minutes to compute [24].

Such time demanding solutions may be suited for planning studies of power systems; however, it is not likely for grid operators to consider its use for time-critical decision making for the operation of the power system. By reducing its computation time, the FOR could be used as a unique interface between grid operators in innovative grid operation and planning tools.

This thesis has the objective to develop a FOR computation method with improved performance compared to the ICPF algorithm of [22]. This is done by redesigning the procedure making use of a fully linear OPF model and by overhauling the search procedure, in order to optimize the number of necessary OPF. The algorithm that is developed and analyzed throughout this thesis follows three main purposes:

- To assess more realistic capability charts of different types of FPU, for which detailed linear models are developed and added to the OPF in the aggregation process.
- To compute the FOR of radial distribution grids with realistic sizes and with significant numbers of FPU.
- Optimize the computational complexity of the algorithm, in order to reduce the processing time and make it conceivable for the algorithm to be considered in fast-paced grid operation concepts.

These aspects are discussed throughout the document following the structure presented next.

1.3 Structure of the Thesis

Three aspects to the analysis of flexibility provision by DER were outlined in [25], which are key to this work. First, economic and technological analysis of flexibility need to be separated, as technologies are more homogenous in a global perspective, while economic frameworks vary between countries. Second, the primary energy provision of DER and the interconnection to the grid need to be analyzed separately, especially as many DER rely on power inverters. Third, the spectrum of analyzed ancillary services and DER units need to be broadened, as most studies just focus on single topics and fail to provide a comprehensive picture. This thesis takes these points into consideration, as the performed analysis of the flexibility of power grids focusses on the technological point of view, with clear distinctions regarding DER units and their grid-coupling technology and an extensive range of DER are considered in the analysis.

In Chapter 2, different aspects and definitions regarding the use of flexibilities for the provision of ancillary services in power systems are given, as well as a comprehensive review on capability chart models of different types of DER.

Chapter 3 provides a comprehensive state-of-the-art review on the many different methods that have been developed to compute the FOR, including analytic, geometric, random sampling, and optimization-based methods. This leads to Chapter 4, which describes a novel linear-OPF method to compute the FOR, as well as different aspects that need to be considered during its computation.

The validation of the model is shown in Chapter 5, including the definition of a set of grid models that are used to showcase the potentials of the developed method. Three different use cases for the application of the proposed model in the operation of distribution grids are shown in Chapter 6. Finally, Chapter 7 provides the closing remarks and the outlook of the thesis.

1.4 Scientific Thesis

The decarbonization and decentralization of power systems are forcing grid operators to explore new alternatives for the supply of flexibility for ancillary services, which are critical to ensure the system stability. As the number of installed generators, storage systems and controllable loads is increasing at the distribution grid level, the cooperation between TSO and DSO needs to be enhanced. This demands for the development of novel methods to share information about the state of the grid, for which the FOR can be one of them. However, if it cannot be assessed fast enough, it is unlikely for it to be considered as valid solution by grid operators. These premises lead to the following scientific thesis:

The FOR of distribution grids can be computed as a result of a linear optimization model, decreasing the computation time, while preserving the accuracy; consequently, offering new opportunities to integrate the FOR concept in grid operation processes.

For this purpose, a comprehensive validation of the proposed linear aggregation method is provided, including a comparison to other available methods. Additionally, three conceivable use cases related to the planning, and specially, to the operation of power systems are given, showing some of the benefits of the reduced computational time.

2 Use of Flexibilities for the Provision of Ancillary Services in Power Systems

2.1 Flexibility in Power Systems

One classical definition of flexibility: "the ability of a power system to change its power output in response to changes in load and generation" [26]. This is what grid operators have been doing for decades with frequency response ancillary services as an example, where the power injection of larger conventional generators is adjusted to absorb sudden changes in the load, e.g. due to the sudden disconnection of a large grid section. However, penetration of RES (at all grid levels), new loads (e.g. electric vehicles, electric heating) and supplementary storage systems (e.g. batteries), the concept needs some rethinking. This has motivated plenty of research towards solutions of grid issues involving the use of flexibility from controllable loads and/or generation. At the same time, it has been shown that the concept of flexibility is very ambiguous, as it can have many different meanings depending on specific use cases. A taxonomy description for the concept of flexibility was given in [27], which divides the concept into three different domains, as shown in Figure 2-1. For each category, some examples are given, however, they are not limited to the ones mentioned here.

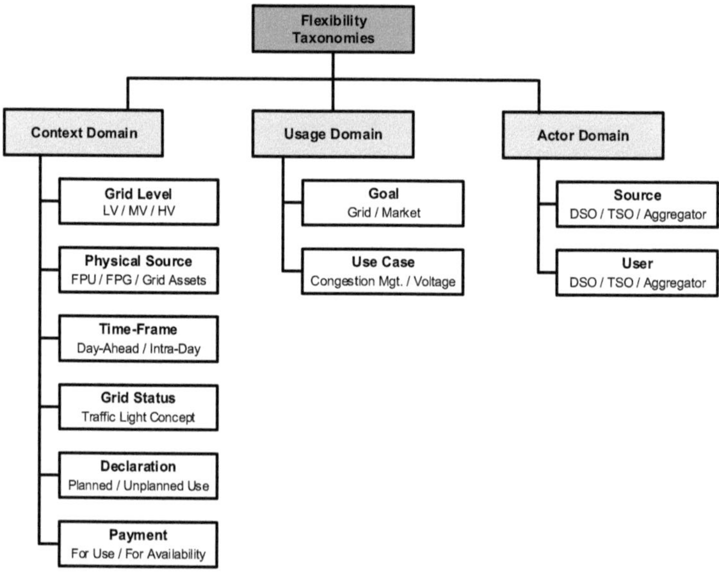

Figure 2-1: Taxonomy for classification of flexibilities. [27]

The concept of flexibility needs to be treated differently according to each specific use case that is being studied, otherwise it is very likely that misunderstandings will happen. On one side, there are some metrics like the loss of load expectation (LOLE), expected energy not served (EENS), loss of largest unit (LLU) or the insufficient ramping resource expectation (IRRE) that allow for analyzing the ability of bulk systems to cope with changing scenarios of RES penetration [26] [28]. These mechanisms allow gaining a broader understanding on what is necessary for a power grid to ensure the stability of its long-term operation, however, they rarely consider the physical limitations of power systems, as grid constraints are normally not integrated into the models. Additionally, these metrics tend to neglect energetic constraints of storage systems. Summarizing, more comprehensive methods to describe flexibility still need to be developed [28].

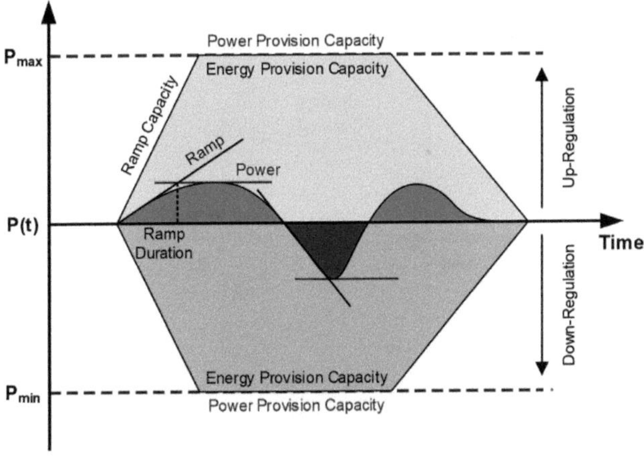

Figure 2-2: Flexibility metrics in power systems operation including power, power ramp-rate, ramp duration and energy. [29]

In [30], the concept of "flexibility envelopes" is defined, describing a set of boundaries that the flexibility provision from RES should fulfill, describing the maximal ramps and the maximal power output. This allows for the quantification of the ability of a storage system to cope with ramping and energy flexibility requirements. In [29], different metrics to describe the envelopes were defined, including maximal power provision, maximal ramp capacity, energy provision capacity and ramp duration. The needed flexibility should be located within the envelope (darker areas in Figure 2-2), otherwise the system would not be able to cope with requirements as either the reaction time (ramp) is slower than necessary, or the power output is too small to satisfy the requirements, or both. Based on [29], the flexibility of a system can be described using the aforementioned parameters for each time-step. One use of this approach is to deal with forecast errors in RES. Some examples on how ramping deficits affect the power system are provided in [31].

Another model to describe flexibility is the "power node", concept that was introduced in [32] and [33]. This model describes, in an abstract way, a node in a power system, which has generation and consumption, as well as a power exchange with other nodes in the grid. This model, also known as "tank model", characterizes the storage capability as well, as illustrated in Figure 2-3. Additional to the power injection, consumption or storage, the model considers all inherent losses related to these processes (e.g. inverter losses, transmission losses), as well as the losses and capacity limits of the storage systems. On one hand, the demand/supply side describes how energy is provided to the node (e.g. wind generators, PV generators) or consumed from the node (e.g. heating, illumination). On the other hand, the node is connected to the grid, which allows to import or export power, in order to preserve the power balance in the node.

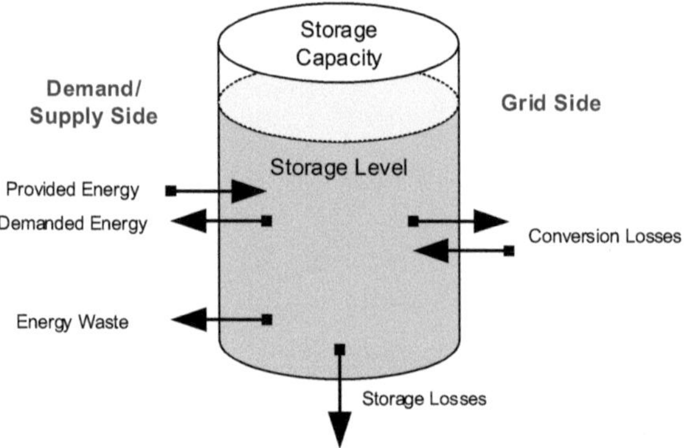

Figure 2-3: Power node concept, linking demand/supply with grid and storage capabilities. [32]

The tank model has been widely adopted in the definition of flexibility market mechanisms in recent years, due to its comprehensive and clear mathematical formulations. The main benefit of this model is that it allows for the description of the time-dependent flexibility of a storage system, as the storage capacity and the state-of-charge are clearly defined. Overall, the tank model allows describing the operational flexibility of the grid. Nevertheless, it only considers active power, completely neglecting the impact of reactive power on the grid.

An attempt to include grid constraints in the quantification of flexibility in power systems, while dealing with uncertainty, results in the "loadability sets" concept. They represent the "allowable ranges of nodal demand that can be met by a given power system" [34]. In [35], a method to compute the "feasible region of loadability" using an (exact) AC power flow approach is provided. In the end, a non-linear

OPF is defined and used to characterize valid operation points of a power system. What is obtained from this model, is a convex polytope[1] representing the combination of operation points of the residual load at the buses that do not violate any grid constraints. This results in an n-polytope, where each dimension describes the operation points of one specific bus. In [36] and [37] the n-polytopes concept is applied in the analysis of multi-area power systems. The loadability set shows which buses have restrictions in the flexibility provision because of grid constraints. In the best scenario, when all buses are considered, the resulting polytope (assuming it exists) defines all valid combinations of operation points in the grid. The model works at a mathematical level; however, it is extremely difficulty to work with an n-dimensional polytope in an algorithm, and even worse at a graphical level. The application of this method to accommodate wind generation was exemplified in [38], though reduced to a small three-bus system, mostly due to the aforesaid restrictions. Even if this method can consider reactive power capabilities in its calculations, most published results and analysis focus primarily on active power.

Capability charts are another method to describe the flexibility potential of power systems. These are "drawn on the complex power plane and define the real and reactive power that may be supplied from a busbar during steady-state operation" [6]. The idea is a century old and was first applied to provide simple graphical illustrations of the non-linear link between voltage and active/reactive power in transmission lines. Since then, it has been used to describe the operation of all kinds of grid elements. In [6], a fully analytical model was provided to compute a capability chart considering all types of constraints in the grid; however, it was not until [8], when an OPF-based systematic approach was proposed, and the concept was revived. Based on [8], a capability chart can be used to describe all valid operation points in the complex power domain when observing from one specific busbar. Different concepts can be included in the calculation, like the volatility (and nowadays also controllability) of RES, usage of storage systems, reactive power compensation, or the simple controllability of conventional generators.

This approach shows some similarities to the loadability concept; however, it works on the premise that the operation of an entire grid section can be represented as a generator with a specific capability chart, which depends on the controllable/volatile load and generation downstream. An equivalent model and its application to the analysis of a distribution grid can be found in [39].

---

[1] Generalization in any number of dimensions of the three-dimensional polyhedron. A polygon is a 2-dimensional polytope. Source: https://en.wikipedia.org/wiki/Polytope

2.2 Flexibility Providing Units (FPU)

The FPU concept defines any kind of controllable generator, storage system or controllable load, which is able to change its point of operation in the complex power domain based on predefined external signals [40]. The concept can be applied to all types of generators, loads and storage systems; however, it focuses on representing DER with the ability to provide flexibility to the power system, while assuming the existence of appropriate control mechanisms. A comprehensive definition of different types of FPU was provided in [29], based on their operational flexibility provision capability, which leads to the summary of DER that could be operated as FPU depicted in Figure 2-4.

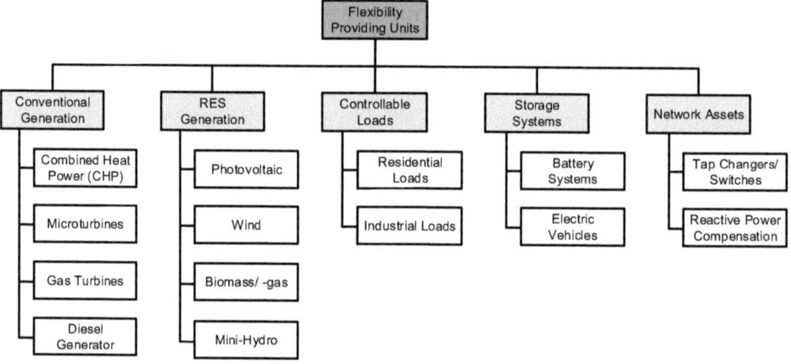

Figure 2-4: Classification of typical DER that can be considered as FPU. [29] [41]

The main characteristic of an FPU is its ability to control its steady-state operation point, otherwise it could not be considered as a flexible asset, as the volatility of the primary energy source should not be mistaken for flexibility. This trait varies among different types of DER, depending mostly on four different factors:

- Primary energy source
- Primary energy converter
- Grid coupling mechanism
- Control mechanism

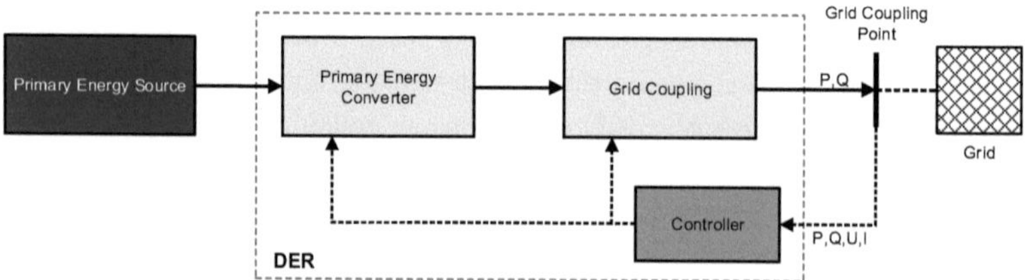

Figure 2-5: Grid interconnection scheme for DER.

The grid interconnection scheme of Figure 2-5 provides an overview of these concepts. The primary energy conversion and the grid coupling depend strongly on the type of DER and the specifications of each device, as both could be controlled to adjust the operation point. Diverse grid coupling technologies can be found in DER, each one with different operational control possibilities:

- Induction Generators (IG)
- Synchronous Generators (SG)
- Doubly-Fed Induction Generators (DFIG)
- Power Inverters

Generation based on traditional electrical machines (e.g. SG, IG) transforms the primary energy source (wind, sun, water, etc.) into kinetic energy by forcing the rotor of the generator to turn. The generator then transforms this kinetic energy into electrical energy, which is directly injected to the power grid. On the other side, converter interface generators (CIG) are coupled to the grid through power inverters. These convert the DC typically generated from chemical processes within photovoltaic arrays, fuel cells or battery storage systems into AC [42]. DFIG generators are a hybrid between both cases, as a power converter is connected to rotor of the IG. This provides the IG with additional controllability, due to the control of the rotor voltage of the IG with help of the power inverter.

In the cases of wind and photovoltaic generation, the volatile primary energy defines the maximal active power output, therefore, directly impacting how much flexibility they can provide. Modern wind generators are equipped with pitch controllers, which in normal operation they would allow limiting the power output, especially in high wind situations, therefore, controlling the primary energy conversion. In deloaded operation, the pitch controller would let the WG provide both positive and negative flexibility, however, this mechanism is not common due to the excessive wear and tear in the blades. Through the operation of a maximum power point tracking (MPPT) in solar PV arrays, the amount of electrical energy extracted from the solar irradiation is optimized. With the remarkable progresses in the development of power electronics in the last decades, novel control mechanisms have been provided for DER, not only allowing the control of active power injection, but also providing them with reactive power support capability, which brings them closer to conventional generation. A power inverter adds additional flexibility capabilities to an FPU; however, inverters are restricted by their own technical limitations [43].

It has not been trouble-free to incorporate RES into traditional ancillary services mechanisms, on one side, due to their volatility, and on the other, because their use cannot be deferred, as the primary energy source cannot be stored, unlike the fuel of conventional generators. Some solutions to this problem have been proposed, for example coupling RES with storage systems or by pooling large numbers of RES. This way, new flexibility options have been developed [25]. The control of geographically dispersed DER has paved the way to the creation of the virtual power plants (VPP) concept [44].

In order to support the grid when voltage or frequency unexpectedly deviates from the set point (i.e. 50 Hz in Europe), DER should be capable of providing instantaneous flexibility to the grid, as any conventional power plant would be expected to do. Based on the grid interconnection point and the size of the DER, specific local control characteristics are required in order to allow the commissioning and grid interconnection. In the case of voltage regulation, the reactive power provision can be controlled using characteristic curves, similar to the ones depicted in Figure 2-6. These can variate among grid operators, as they need to be adapted to local grid requirements. For example, in Germany, the interconnection of DER into LV grids is regulated by VDE-AR-N 4105 [45], setting specific reactive power requirements. Similar case for the MV level with VDE-AR-N 4110 [46].

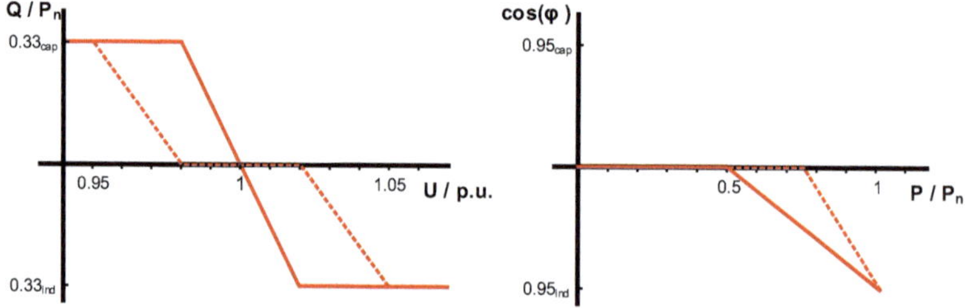

Figure 2-6: Examples of characteristic curves for DG. (left) Q(U), (right) cosφ(P). [47]

Besides generators and storage systems, also consumers can be flexible. Different types of customers can have entirely different consumption patterns, based on the types of electrical devices connected and the way they are used. A summary of different types of loads and some examples for each are given below:

- **Residential**: Houses, apartment blocks
- **Commercial**: Local shops, shopping centers
- **Industrial**: Factories, lumber, mining, warehouses, ports, airports
- **Agriculture**: Farms, both arable and pastoral
- **Municipal**: Public lightning, traffic lights, water supply
- **Traction:** Public transport, trains, electric cars

As the main objective of power systems is to serve loads 100% of the time, the development of flexibility options for loads has been gradual, yet different concepts have been developed over time. Diverse demand side integration (DSI) concepts have been proposed, ranging from large industrial customers all the way down to individual appliances within a household. This includes the disconnection of large loads at the MV/HV level for redispatch or frequency support (5MW for MV and 50MW for HV in Germany)[2], as well as the load profile optimization of households [48]. Aggregators have been established to harvest flexibility from loads by applying DSI methods to large numbers of household appliances and/or by curtailing/shedding industrial loads [49].

2.3 Vertical Provision of Flexibility for Ancillary Services

The increasing penetration of DER in detriment of the traditional large-scale centralized power plants (e.g. coal, nuclear) is changing the way power systems are operated. One worrying aspect of this change is the sustainability of ordinary grid ancillary services, which are essential to maintain a stable system operation[3]:

- Congestion management
- Frequency control (primary, secondary, tertiary)
- Steady-state voltage control
- System balancing
- Reduction of technical losses
- Short circuit current
- Inertial frequency response

With the inclusion of DER, the increased controllability of loads and the increased overall flexibility in the power systems, novel ancillary services are already appearing or are expected to emerge in the coming years [50] [51]:

- Fast voltage control (primary, secondary and tertiary)
- Black start capability
- Fault ride-through capability (FRT)
- Power quality
- Inertia emulation
- Ramping control

So far, most frequency-based ancillary services (including balancing) are performed exclusively at the TSO level, as it involves the use of large conventional generators. These services are in most cases market-based and have been well-structured for years. Congestion management and overall steady-state voltage

---

[2] "Abschaltbare Lasten", www.bundesnetzagentur.de
[3] "EUETS Market Glossary", www.emissions-euets.com/internal-electricity-market-glossary

control are steered with bilateral agreements with power plant operators or the operation of reactive power compensation units. This is expected to change with the penetration of DER, as the universe of will FPU increase dramatically, forcing the development of new methods to accommodate the new requirements.

In [2], a comprehensive presentation of the changing energy landscape was provided, including multi-energy systems, microgrids, electric vehicle charging stations, demand side management, and virtual power plants. All these new actors need to contribute to the overall grid stability, requiring entirely new control strategies, at local, centralized, and distributed level. A more precise explanation on these control strategies can be found in [47].

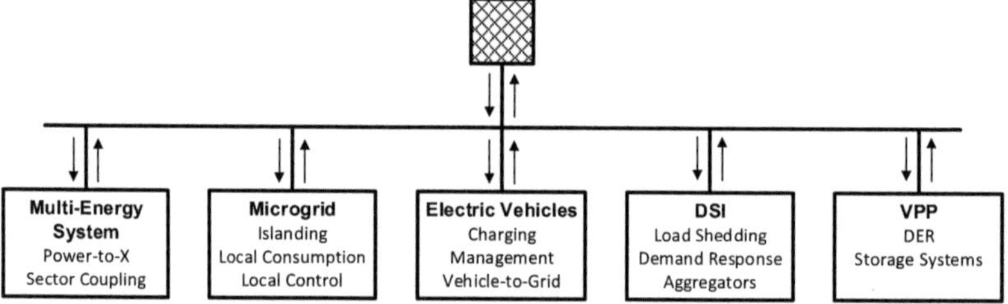

Figure 2-7: *Evolving energy landscape with bidirectional power flows due to novel smart grid concepts. [2]*

A widely discussed approach involves increasing the cooperation between TSOs and DSOs, as most new connections of DER are located within DSO grids. In [52] and [53], five novel TSO-DSO cooperation concepts for the provision of ancillary services across voltage levels are proposed. The proposed roles have a more market-oriented focus, in order to allow for the inclusion of so-called flexibility markets into the operation of grid operators (e.g. USEF[4], FLECH [54]). These aspects involve the vertical provision of flexibility, which is a topic that has been discussed in, inter alia, [7], [55], [56], [57], and [58]. They describe how the flexibility provided from FPU located at lower voltage levels could be used to provide ancillary services to the upper parts of the grid. In [59], a methodology to meet specific power transfer standards that allow providing the required ancillary services is described. These publications join the concepts of aggregated capability charts and provision of ancillary services, one of the main topics covered in this thesis.

---

[4] https://www.usef.energy/

2.4 Capability Charts of FPU

Every device connected to the grid, e.g. loads, generators, or storage systems, either generates or consumes (in steady-state) a certain amount of active and reactive power. This operation point may change over time due many factors, i.e. local or global control strategies, changes in primary energy sources, disconnection from the grid. Regardless of what causes the adjustment, the operation point should at all times be enclosed within certain limits, which defines stable operating states of the specific asset. These boundaries are defined as the capability chart of the component, as seen from the grid perspective. The operation points outside of the boundaries are either unreachable, i.e. a wind generator (WG) cannot inject nominal power with less than nominal wind speed, or cause internal problems in the device, i.e. thermal failure of the components due to overcurrent.

The capability charts concept has been closely associated to traditional SG [60]. The operation of a SG has been studied in depth for over a century now, allowing to create a very detailed graphical description of its capability chart, together with the correspondent analytical equations that define the characteristic operation points. One important aspect of capability charts, is that they can describe the reactive power capability of DER and its linking to the active power provision. This chapter describes the capability charts of typical DER and the extension of the concept to represent the FOR of specific grid sections [9].

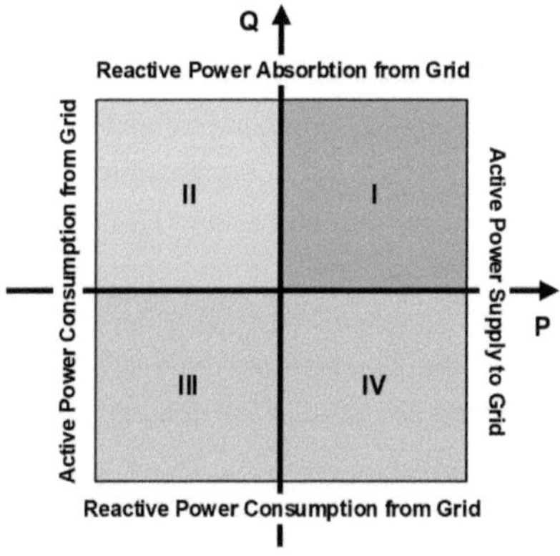

Figure 2-8: Four-quadrant grid operation chart based on generation reference system.

2.4.1 Definition of Reference Convention

The capability chart depicts the active and reactive power operation points of a generator, load or storage system (which can inject and absorb power). The complex power $\underline{S} = P + j \cdot Q$, $P, Q \in \mathbb{R}$ connects both terms. Based on this, a four-quadrant operation of DER can be described when required, based on the sign of P and Q, as shown in Figure 2-8. This work adopts a generator-reference arrow system convention (generation/injection is positive), which is used in this thesis.

2.4.2 Capability Charts of Different Types of DER

Over time, capability charts of different types of devices have been developed as well, e.g. HVDC systems, wind generators, controllable loads, storage systems [60]. In some cases, it has been achieved through analytical approaches, while others have relied on numerical simulation models or empirical models based on measurements. As stated in [25], "the technological capabilities should be analyzed separately for the grid-coupling converter and the whole DER unit". On one side, the PQ provision of any DER unit is limited by different technical limitations, which define the capability chart of the unit. On the other side, the device that interconnects the DER to the grid, e.g. a power converter, possesses its own limitations, additional to the restrictions of the DER unit. In case of inverter-based technologies, it has been observed that the voltage at the interconnection point to the grid can have a large impact on the profile and extension of the capability chart. This subchapter describes the capability charts of different types of DER.

2.4.2.1   Synchronous Generators (SG)

Synchronous generators have been a key component in power systems since the very beginning and are currently being used in different types of DER. Their operation has been analyzed for many years, therefore, the construction of their capability charts is well known and a can be represented through a defined equations system. The main limitations in the operation of a SG are the stator and rotor currents, as well as the rated current of the machine. The interconnection to the power grid imposes additional constraints to the operation of the SG, resulting in a voltage stability limitation, as described in [6], [61], and [62]. Other factors can limit the capability chart of the SG, one example is the case described in [61], which details the impact of the pressure of the cooling gas (H_2 in the case of [61]).

The typical capability chart of a SG is shown in Figure 2-9, where the shaded area corresponds to the FOR, constrained by different construction limitations of the machine (e.g. rotor, turbine, windings, terminals voltage) [6] [61]. The FOR can change according to the technology behind the SG, e.g. [63], where a detailed analysis of the capability chart of combined cycle power plants was provided. The voltage stability margins may cause the capability chart to become non-convex, yet in most OPF analysis, it is simply assumed as convex, or even oversimplified as a rectangle. This effect is observed in other types of generators as well. The studies of capability charts applied to SG have evolved to the point that they can be assessed dynamically based on measurements of the machine and displayed online, with the possibility to change parameters, as is presented in [5].

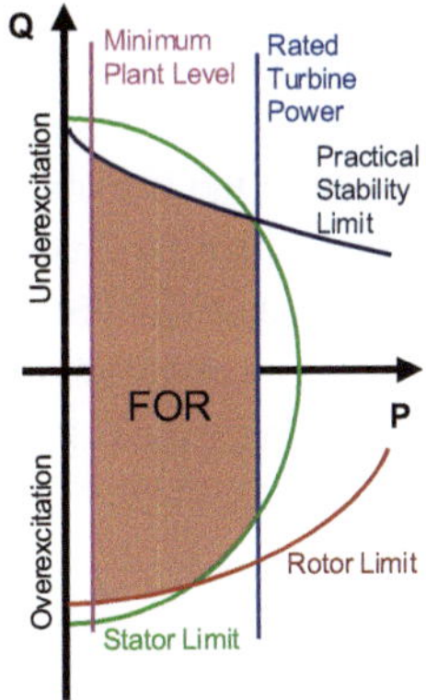

Figure 2-9: Capability chart of a synchronous generator [6] [61].

2.4.2.2   Photovoltaic Generators (PV)

Nowadays, photovoltaic generation can be found in every voltage level of the power grid, as its modular design allows for scaling its size from a few kW up to several MW. At the distribution level, PV was originally designed to operate with unity power factor, although this is changing with time, as reactive power provision could be required from them. An array of PV panels is connected to the grid through a DC/AC power converter, which also adds the reactive power capability [64]. Different interconnection topologies of PV are shown in Figure 2-10. The PV arrays can be connected symmetrical or unsymmetrical to the grid, where the transformation stage can be done all together or split per phase [65].

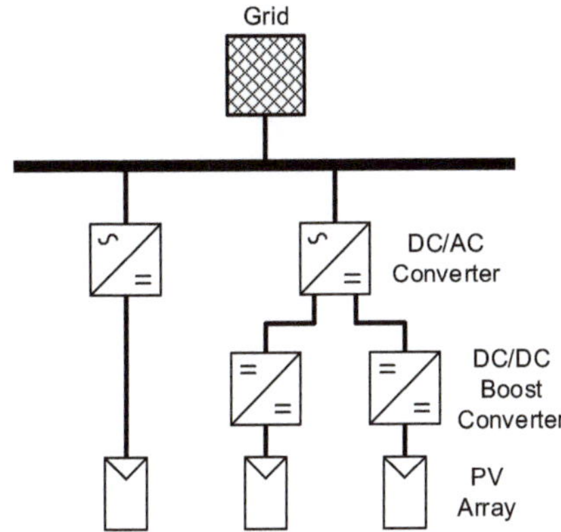

Figure 2-10: Different connection topologies of PV arrays to the power grid. [64] [65]

The capability chart of a PV generator is mostly defined by the solar irradiance, the MPPT and the characteristics of the power converter [66]. The power converter is usually rated to support the maximal apparent power of the PV generator, which results in reactive power provision capability at zero active power injection levels. Unless the inverter is oversized, no reactive power can be provided at rated active power injection. The voltage at the terminals of the power converter may limit the reactive power provision [67] [68]. Additional aspects, like the operation of the MPPT, of the voltage source converter (VSC) PWM or the ambient temperature can put additional limits to the capability chart [68] [69]. Different capability charts for PV generators are shown in Figure 2-11.

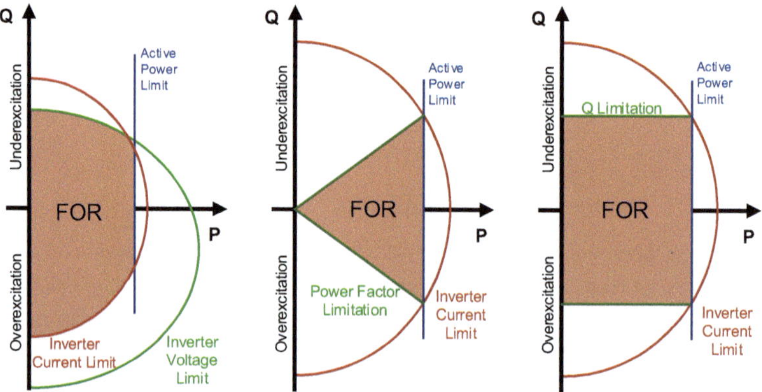

Figure 2-11: PV capability charts considering different control mechanisms. (left) Power inverter technical characteristics [67] [68], (center) Power factor limitation [70], (right) Reactive power limitation [70].

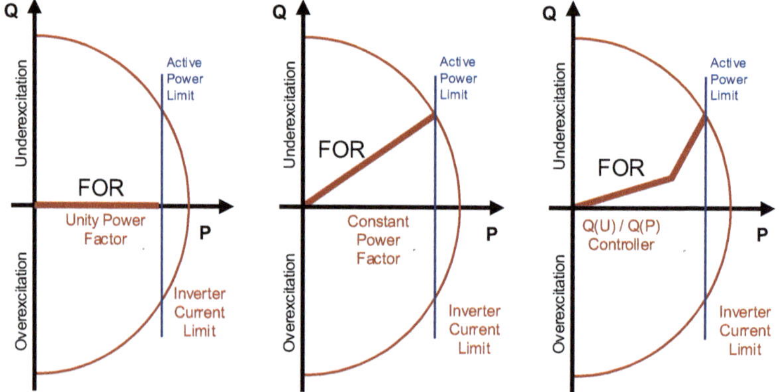

Figure 2-12: PV capability charts considering limited reactive power control. (left) Unity power factor, (center) Constant power factor, (right) Q(U)/Q(P) controller.

In general, PV generators are expected to have a FOR that can be easily modelled using convex polygons, including the simple cases, where no reactive power control is available or the PV is operated with a constant power factor (e.g. linear FOR characteristics in Figure 2-12). Additional modelling efforts are required in cases where the PV has a local $Q(u)$ or $P(\cos(\varphi))$ controller [47]. In [71], the operation of a real PV was analyzed, and based on measurements a non-convex capability chart was constructed. The normal operation of the PV was analyzed, where the power factor setting of the inverter was self-regulated by a closed-loop controller. This does not necessarily show all possible operation points that the PV could achieve if the inverter was controlled differently, just the ones that it achieves with the use of this particular controller configuration. Even though, it should be noted that the capability chart concept may include operation points that the device would not necessarily reach in everyday operation.

2.4.2.3   Wind Generators (WG)

Wind generators were designed to be controllable from the very beginning, either to allow an optimal harvesting of wind energy, to provide ancillary services to the power grid or to protect the turbine from very high wind speeds. Nowadays they are expected to inject not only active power, but also reactive power, in order to provide voltage support and FRT capability. Wind turbines can be operated with fixed- or variable speed, although the second type is the most commonly found nowadays. A comprehensive description of this technology can be found in [72]. Four traditional WG topologies (defined as Type 1 to 4) are found in literature, each depicting a different combination of generator and power converter [73]. These topologies provide different capabilities to the WG for injecting active and provide reactive power support to the grid. Wind generators of Type 3 (doubly-fed induction generator; DFIG) and 4 (full converter wind turbine; FCWT) are the most commonly found today, therefore, their flexibility provision capability is detailed here (Figure 2-13 and Figure 2-14). A summary of different capability charts that have been derived for DFIG is shown in Figure 2-15.

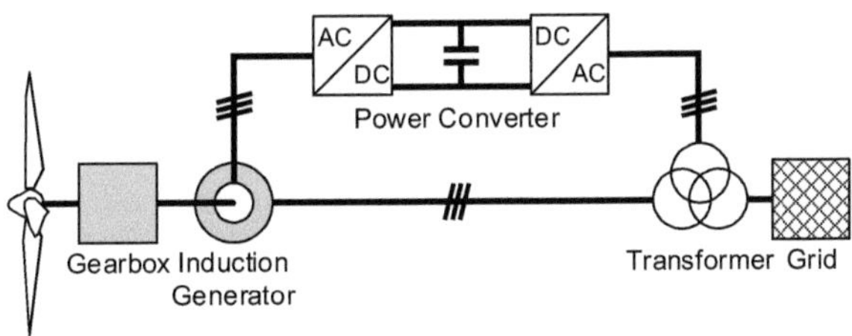

Figure 2-13: Type 3 wind generator - Doubly-Fed Induction Generator (DFIG). [72] [73]

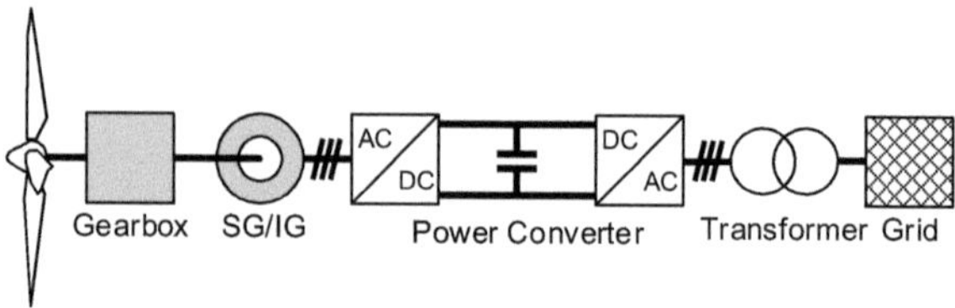

Figure 2-14: Type 4 wind generator - Full Power Converter Wind Turbine (FCWT)). [72] [73]

- **DFIG:** The capability chart is defined by the active power provision from the stator and the reactive power based on the frequency of the rotor voltage. Both values are coupled through the slip s [74]. The power converter connected to the rotor of the IG would allow the control of the rotor frequency between 0,75-1,25 p.u. Therefore, the compact power converter allows for the control of around 30% of the rated power, contributing to a two-quadrant operation of the DFIG. The voltage at the PCC certainly impacts the capability chart, as it limits the reactive power capability of the DFIG. The impact of s, the MPPT operation, and the voltage at the PCC. The operation of DFIG is analyzed in detailed in [72], [75], [76], [77], [78], and [79]. Reactive power provision of DFIG is analyzed in [80] and [81]. Different DFIG capability charts are shown in Figure 2-15.
- **FCWT:** The generator is decoupled of the grid by the power converter, hence the FOR of the FCWT depends mostly on the converter properties. The converter has usually a larger rated power that the generator, which allows for reactive power provision at rated active power output. In [82] and [83], analytical models for the converter voltage limits were derived. These equations describe how the voltage at the PCC can limit the reactive power provision of the FCWT. Moreover, power electronics tend to have problems providing reactive power at low active power injection levels, issue that can be mitigated with the addition of a STATCOM, either at the busbar or in the DC loop of the inverter. These effects in the FOR are shown in Figure 2-16.

Each wind turbine type owns a unique set of features, which can vary drastically among manufacturers, making it hard to derive a single capability chart to represent every type of wind turbine. Several publications have focused on validating the static capability charts with more detailed models of wind turbines, e.g. [84], [85], [86] or [87]. From these, two main conclusions can be drawn. First, the steady-state operation of the wind generators can be modelled in the complex power space and can be approximated by a convex irregular polygon. Second, the voltage at the PCC has a strong influence on the reactive power capability of

WG and can result in non-convex shapes, although these can be approximated into convex shapes without losing much generality. However, this is in some cases strongly dependent on the state of the grid and the control of the WG [88].

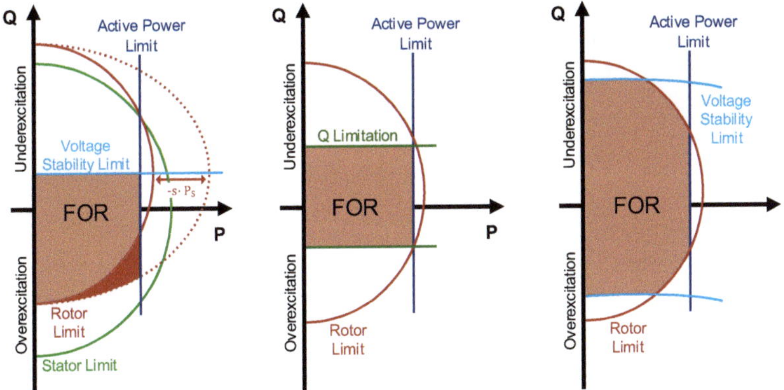

Figure 2-15: Capability charts of DFIG WG. (left) Impact of slip of induction generator [74] [89], (center) Reactive power limits [90], (right) Impact of rotor voltage stability limits [77].

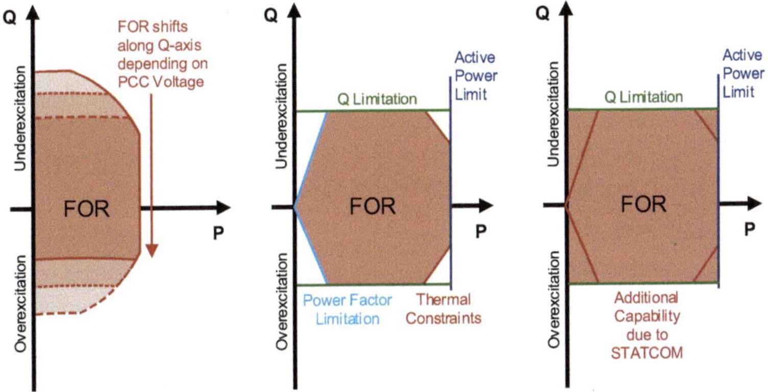

Figure 2-16: Capability charts of FCWT. (left) Impact of PCC voltage [82], (center) Reactive power restrictions of power inverter [85] [91], (right) Added capability of STATCOM [85] [91].

2.4.2.4   Energy Storage Systems (ESS)

Storage systems have become key components in ADN, especially when it comes to assisting the control of RES volatility. ESS work by transforming electrical energy into a variety of other types of energy, which can be stored. A summary of different methods to store energy is shown in Figure 2-17, where a categorization based on the form of energy is provided. Comprehensive summaries of ESS technologies can be found in [92] and [93].

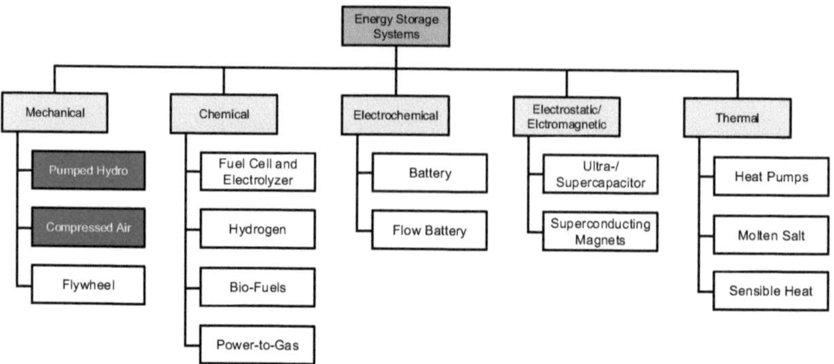

Figure 2-17: Different energy storage systems, categorized by form of energy (the dark shaded technologies cannot be considered as DER due to placement constraints). [92]

Many types of storage technologies are in use, yet not all of them can be labelled as DER. For example, pumped hydro accounts for the largest storage capacity worldwide, yet the technology is limited by geographical constraints and large area requirements, therefore, does not fit in the definition of DER. The same happens to some compressed air plants (e.g. using mining shafts), or hydrogen storage technologies, which require great volumes to store large-scale amounts of energy, due to properties of H_2. Figure 2-17 characterizes different storage technologies, and displays which are suited to operate as DER. This means, that those technologies are more likely to be found connected at the DSO level. However, plenty of research is being performed to scale down different types of storage technologies, meaning that Figure 2-17 could change in the near future.

The interconnection of ESS to the power grid depends on the type of technology. Chemical, electrochemical and electrostatic or electromagnetic systems require an AC/DC power conditioning system to convert the current when charging and discharging. Mechanical and thermal units use AC electrical machines (motors/generators), therefore, they could be directly connected to the grid. A review on different interconnection topologies of ESS into distribution grids is given in [94], considering DC and AC technologies and specifying the required power converters for each case. In contrast to other types of generation technologies, there is not much information available regarding the capability charts of ESS, besides their ability to absorb and inject active power. As most distributed ESS are connected to the grid using power electronics, they could provide reactive power support as well, depending on the adopted control system. However, in most cases, the capability chart is described as a circle, defined by the current limitations of the power inverter (e.g. [95], [96]), yet as shown in [43], the converter adds additional constraints, meaning that each ESS should be modelled separately.

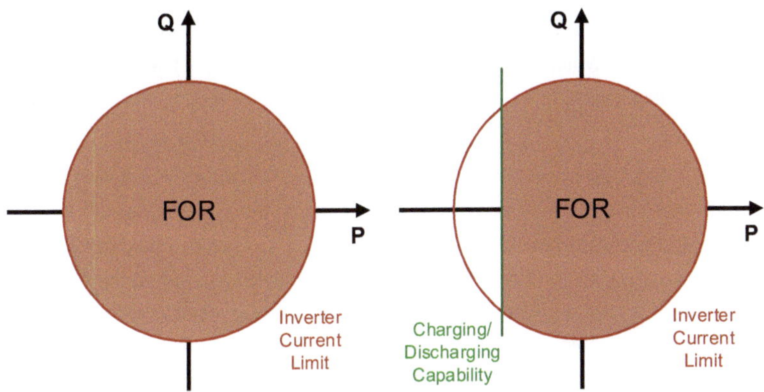

Figure 2-18: ESS capability charts. (left) Full converter rating [95] [96], (right) Different charge/discharge power limits [43].

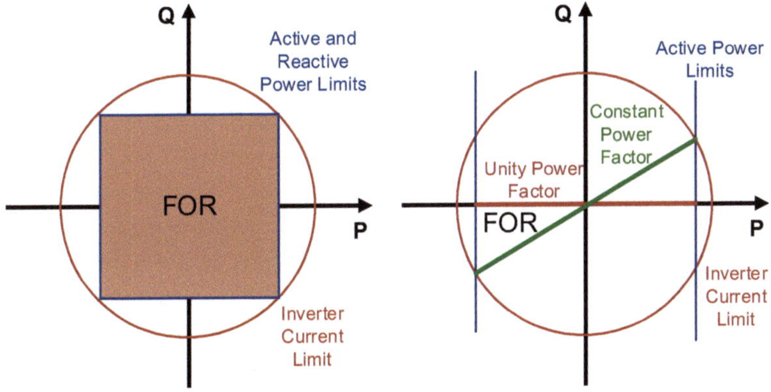

Figure 2-19: ESS capability charts. (left) Simplified rectangular approximation [41], (right) Operation with unity or constant power factor.

2.4.2.5   Controllable Loads

The behavior of power consumers is stochastic and has historically been seen as passive, meaning that a reduced number of control mechanisms have focused on the control of loads. So far, power systems have been operated in a way that generation tries to match the behavior of the loads as best as it can, meaning that there always needs to be more generation available than the expected load for a given time. Load shedding has always been a useful mechanism for grid operators to overcome critical stability issues in the system (e.g. large frequency deviations), yet it involves leaving hundreds if not thousands of customers without electricity for a large amount of time. This can be limited to a few very large industrial customers as well. As an example, in Germany, a minimal of 50 MW should be curtailable in the HV, while in the MV only 5MW are necessary[5]. These quantities do not necessarily need to be provided by single assets, as they can be aggregated.

In recent years, many different methods to model and control an array of flexible loads have been proposed, yet most of them have focused on regulating thermostatically controlled loads, e.g. heating, ventilating and air conditioning (HVAC), water heaters, or refrigeration systems [97]. These flexibilities could have a noteworthy impact on the grid by themselves, especially at the industrial level. With the arrival of smart appliances inside households everywhere, new sources of flexibility are available as well, e.g. washing machines, tumble dryers, or dishwashers. The impact on the grid stability of single devices is generally insignificant, however, when aggregating large numbers of these devices together, the impact can become meaningful [49].

Electrical vehicles are contemporary controllable loads, which draw plenty of energy from the grid, especially considering fast-charging solutions. An interesting aspect of EV is that their charging profile can be controlled, either by switching on/off, or by adapting the charging power. In vehicle-to-grid (V2G) concepts, they act as moving ESS, providing temporal and spatial flexibility. As EV charging stations are not yet properly standardized, their interconnection to the grid and their controllability can differ drastically between producers. A critical aspect is the stage at which the AC/DC conversion is performed. In general, just a discrete change in the charging power is possible (i.e. either charge or not charge), while just a few charging stations would allow controlling the charging power, if the power inverter allows it. In most cases charging stations are operated with constant power factor, even if the power inverter would allow reactive power control.

In most cases, a load can just be turned on or off, and its usage can be deferred. From the discrete behavior of a single unit, a continuous linear approximation of the aggregated behavior of many loads can be obtained. This was proposed in [97], where the capability chart of thermostatically controlled loads is derived. The use case analyzed in [49] focuses on providing flexibility by shifting the aggregated discrete operation of white goods over time. In [98], the capability chart of an industrial load is constructed, based on measurements. When considering load profiles, usually for grid planning and operation purposes, a constant power factor is generally assumed and these measurements validate this assumption.

---

[5] Verordnung über Vereinbarungen zu abschaltbaren Lasten (AbLaV), 2016

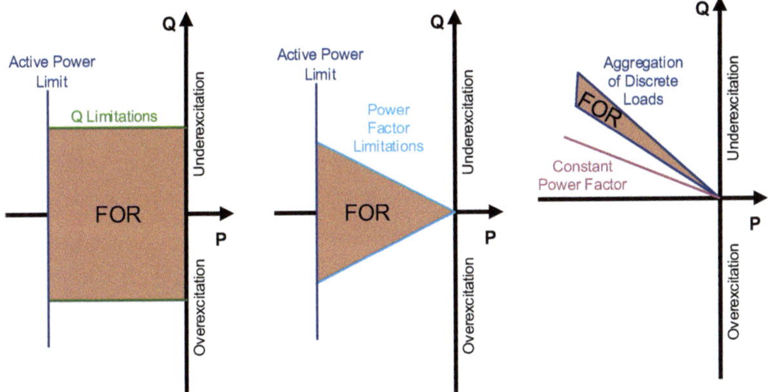

Figure 2-20: Capability charts of controllable loads. (left) Simple rectangular boundaries for loads with highly variable reactive power, (center) Power factor limitations, (right) Constant power factor and aggregated discrete flexibilities [97] [98].

A set of capability charts representing the operation of controllable loads is provided in Figure 2-20. This allows for representing the operation of some types of loads, but not all of them, as the universe of devices supplied by power grids is enormous. The capability charts of EV, including V2G concepts, can be derived from the functionality of storage systems, as were detailed in the previous chapter. In this work, the flexible operation of loads is limited to the given capability charts.

2.4.3 Grid Codes Requirements

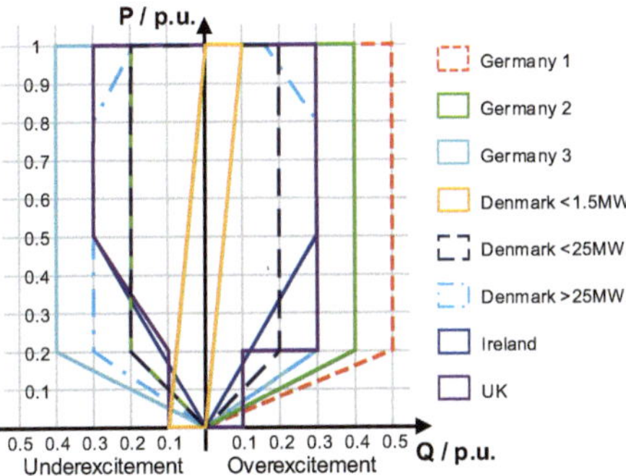

Figure 2-21: PQ grid code requirements in selected European countries. [99] [100]

Grid operators have defined norms for the interconnection of loads and generators to their grids. Grid codes define, among other aspects, active and reactive power requirements for DER, usually in the form of capability charts. The diagrams focus on limiting the reactive power injection/absorption of the device, by setting strict Q-limits or limiting the power factor, which DER controllers should respect at all times. There is a consensus among grid operators on the necessity of having grid codes; however, the specifications can vary significantly between two different grid codes. This can happen because of the different grid characteristics, however, in many cases this happens simply due to historical reasons. Larger problems arise when the grid codes do not meet modern requirements or expectations.

In [101], a review on grid code requirements in the USA and Europe is given, while a comprehensive analysis of European grid codes is provided in [99], from which a more recent update is given in [100]. The strong penetration of WG, PV, ESS, EV and controllable loads in the grids has forced many grid operators and regulating organisms to update the grid codes in the last years. One special focus has been given to the reactive power provision by DER (e.g. the evolution of the IEEE standard on interconnection of DER [102]). The grid codes tend to generalize their requirements to all types of DER connected.

Usually there are inconsistencies in what the grid operator requests from interconnected devices in the power grid, compared to what these devices can actually perform. One example is given in [84], where operating limits of fixed- and variable speed wind generators are compared to a 0.95 power factor limit (both lag and lead). The results show that the analyzed fixed-speed WG could only comply with the limitations imposed by the grid operator when a capacitor bank is added. Similar results were shown in [86] for a wind park in Egypt, describing the necessity to have an additional STATCOM capabilities to fulfill the grid code requirements. Grid codes provide general guidelines on the expected capability charts of DER, which cannot be ensured to be satisfied by every device. When there is no information about the true capability of a DER, the local grid code requirements could be used to fill this gap and adopt the expected model.

3　Feasible Operation Region of Active Distribution Networks

The integration of DER in the MV and LV grid levels started with limited controllability, because of technological (i.e. outdated Type 1 fixed-speed WG have no external control mechanisms) and economical limitations (i.e. the cost of power inverter technologies has strongly decreased in the last decade). This is changing in the present years, as controllability from DER is increasing, both in quantity and quality. As the number of grid-connected units increases, regardless of the type of interconnection (power inverter or electrical machine), the complexity of system control strategies increases accordingly. It is not trivial to develop mechanisms that can handle distribution grids with a large number of feeders and controllable devices. In order to alleviate the task, the development of algorithms to produce aggregated information becomes necessary, so that the computational complexity of the algorithms, usually in the form of optimization problems, can be reduced.

The first suggestion of using the classical capability charts of SG (as discussed in Chapter 2.4) to describe the operation of power systems in general was provided in [6], in 1987. There, the following definition was provided: "The two-dimensional capability chart associated with a particular busbar can be regarded as being a single slice of an overall $2n$ dimensional capability chart for the n busbars that make up the general power system. This overall capability chart describes all of the combinations of complex power simultaneously available from the n busbars". This is the first known description of what would three decades later became known as the feasible operation region, or the FOR [7] [14].

In [6], not only the application of capability charts to describe in an aggregated way the operation of the grid was conceived, but it also offered the first systematic approach on how to compute the FOR. This process was usually performed in those days with analytical and graphical representations. By means of a contour tracing method, a sequence of loci on the power complex plane are constructed, allowing the application of the algorithm for general use in power systems. One novelty of the approach is the fact that the grid constraints are also included in the calculation. The effect of including grid constraints in the calculation, instead of assuming a simple "copper plate" approach has been comprehensively explained in [103].

The original method of [6] proved to be effective in the computation of the FOR, yet computationally inefficient, mainly due to the mixed-integer non-linear optimization problem that is formulated. Since then, many different researchers have dedicated their time to develop novel methods to perform this task. The main objective has been to improve the computational efficiency of the algorithms.

The objective of this chapter is to provide the reader with a state-of-the-art overview of all the different published methodologies to compute the FOR of power systems. By the time this thesis was being finalized, a comprehensive review on FOR computation techniques was published in [104], mostly supporting the contents review.

3.1 Feasible Operating Region (FOR)

Different analogies between the operation of power grids and the operation of SG were made in [6], [9], and [105]. Since then, the capability chart concept was adopted to be used not only for analyzing the operation of SG connected to the grid, but also for representing the operation of sections of the grid itself, considering the impact of these SG connected to it. In [10], it was proposed to use the capability chart concept to analyze the impact of the grid interconnection of a new wind park, showing that the method could be applied to multi-bus grids. Later, the approach was used to show the permissible ranges of power exchange between two grids, taking the relationship between active and reactive power into account [11]. The application of the concept into a real power grid was shown in [24], giving insight on the potential application of the concept in grid planning tasks with TSO-DSO cooperation in focus.

In [7], the differentiation of the capability chart of single DER units with the one representing power systems was proposed, hence the capability chart of an ADN became known as the FOR. This is defined by a set of valid interconnection complex power flows (IPF) of the ADN, which represent the changes in the imported/exported power when FPU change their operation point restricted by their own capability charts. The FOR concept can be applied to represent valid operation points of a grid section, however, it can be related to the virtual power plant (VPP) concept (e.g. [44] and [106]) or to microgrid concepts (e.g. [41]) as well.

It was noted in [14], that the FOR concept was not enough to represent the flexibility potential of a grid, as the time that an FPU requires to change its operation point is not taken into consideration in the capability charts concept. This is can be of special interest when considering the problems envisioned with the so known "duck curve" described in [107] for the California ISO, which describes the necessity of flexibility with extreme ramping capability to counter the steep changes in the residual load as solar plants suddenly start and stop injecting power into the grid at sunrise and sundown. As some FPU can react faster than others, the FXOR concept was proposed by [14]. The FXOR is contained within the FOR, as it is a subset of valid grid operation points which can be provided by FPU with similar reaction time. Operation points located far from the initial value will require flexibility from a larger number of FPU and this may include slower devices. This means that the reaction time to a required change in the operation point will be slower, limiting the usage of that flexibility for specific ancillary services. For example, an FPU with slow reaction time could not be able to fulfill the prequalification requisites for participation in frequency control mechanisms. The FXOR concept is also suitable to define the difference between the technically feasible (time-independent) and the currently available (time-dependent) flexibility of an FPU, as illustrated in Figure 3-1.

The structure of the FOR and FXOR concepts are essentially similar, i.e. they are both defined as convex polygons, therefore, for the sake of simplicity, this thesis will refer to both as the FOR, without losing generality. The FOR is usually the result of the aggregation algorithm, while a capability chart represents the operation boundaries of FPU, which are defined either by technical or operational (i.e. flexibility provision) limitations. However, in some cases the capability chart of an FPU is also described as the FOR.

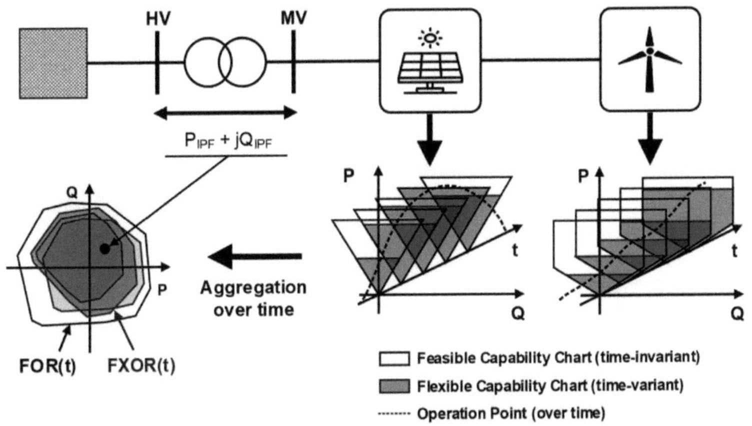

Figure 3-1: Schematic representation of the FOR and FXOR notions applied to an ADN. [108]

One additional use of the FOR jargon is made in this thesis. In a multi-stage concept, where flexibility is aggregated bottom-up, beginning from the LV grids up to the HV grids, the FOR concept takes a supplementary meaning. A grid section that can provide flexibility, due to the combined flexibility of the local FPU, is defined a flexibility providing grid (FPG). Therefore, the FOR of this grid section defines the capability chart of the FPG, in order to allow its application in the FOR computation of the overlayered grid. This mechanism is applied for TSO-DSO coordination in [56] and for microgrids analysis in [41].

3.2   Approaches to Compute the FOR of Distribution Grids

3.2.1 Analytic Approaches

The study of the operational stability margins of power systems has been an important research topic for many years, which has resulted in a number of analytical representations of the power system. Some of these approaches look for the limitations in the active and reactive power injection at specific buses before the grid reaches the voltage limits or the thermal limits of the power lines. The analytical analysis of power systems allows to construct the capability chart of a synchronous generator interconnected to the grid, as shown in [10], [109] and [110].

This analysis is illustrated using the simple two-bus model shown in Figure 3-2, which has a synchronous generator connected to one bus (with an injection of power defined as $\underline{S_G} = P_G + Q_G$), and the other bus is defined as the slack. Both buses are connected through a power line with complex impedance $\underline{Z}$. Through a mesh analysis, applying the second Kirchhoff law, the complex current $\underline{I}$ transferred through the power line can be calculated, resulting in (3-1). This equation is rewritten in terms of the complex generator bus voltage $\underline{U}$ divided in real and imaginary parts, resulting in (3-2) and (3-3)[6]. A detailed explanation of the derivation of these equations can be found in Appendix B. The maximum allowable power line current is I_{th}, while U_{min} and U_{max} describe the voltage magnitude limits at the PQ bus. The operational limits of the generator are given by (3-6) and (3-7).

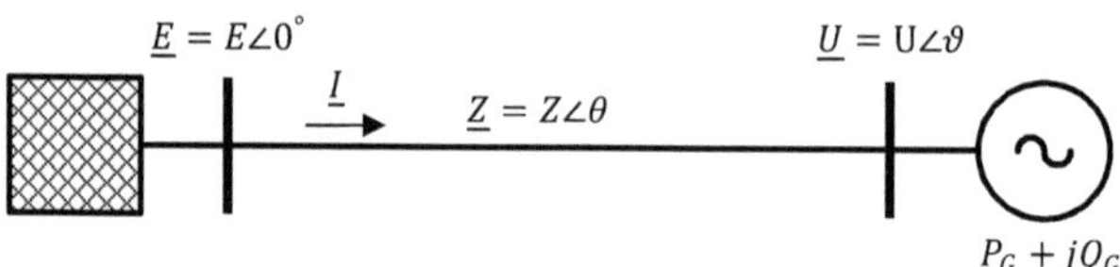

Figure 3-2: Simplified two-bus power system. [10]

The operation point $P_G + jQ_G$ of the generator, represented as a PQ-type bus in the two-bus model, is described in (3-2) and (3-3) in the rotated complex generator voltage reference axis, as shown in Figure 3-3 (left side). The operation point is bounded by $P_{Gmin} \leq P_G \leq P_{Gmax}$ and $Q_{Gmin} \leq Q_G \leq Q_{Gmax}$. The power line thermal limit I_{th} bounds the maximal transmitted current according to (3-4), while the voltage stability limit of the grid is defined as $U \cdot cos(\vartheta) = 0.5 \cdot E$ [10]. Both constraints can be represented in reference to the generator voltage, including the voltage constraints of (3-5), as depicted in Figure 3-3 (right side).

---

[6] Power line losses are neglected in these equations.

$$E = \underline{I} \cdot \underline{Z} + \underline{U} = \frac{S_G^*}{\underline{E}^*} \cdot \underline{Z} + \underline{U} \tag{3-1}$$

$$U \cdot cos(\theta - \vartheta) = E \cdot cos(\theta) - \frac{P_G \cdot Z}{E} \tag{3-2}$$

$$U \cdot sin(\theta - \vartheta) = E \cdot sin(\theta) - \frac{Q_G \cdot Z}{E} \tag{3-3}$$

$$|\underline{I}| \leq I_{th} \tag{3-4}$$

$$U_{min} \leq u \leq U_{max} \tag{3-5}$$

$$P_{Gmin} \leq P_G \leq P_{Gmax} \tag{3-6}$$

$$Q_{Gmin} \leq Q_G \leq Q_{Gmax} \tag{3-7}$$

With: $\underline{E}, E\angle0°$ Complex voltage at slack bus (with reference angle)

$\underline{I}$ Power line complex current

I_{th} Power line thermal limit

$\underline{Z}, Z\angle\theta$ Power line complex impedance

$\underline{U}, U\angle\vartheta$ Complex bus voltage

U_{min}, U_{max} Bus voltage magnitude min. and max. boundaries

P_G, Q_G Generator operation point, active and reactive power

P_{Gmin}, P_{Gmax} Generator active power operational limits

Q_{Gmin}, Q_{Gmax} Generator reactive power operational limits

As the described constraints are all referred to the same reference frame, the capability chart results from the intersection of the areas delimited by the constraints, which represent the valid operation points of the generator. The result of this example is shown in Figure 3-4, which can be translated to the PQ cartesian space by applying the transformations of (3-8) and (3-9). It is noticeable that the resulting capability chart computed in this way can be non-convex, due to the impact of the generator voltage in the boundaries of the operation points.

$$P = \frac{x \cdot E - (x^2 + y^2)}{Z} \cdot cos(\theta) - \frac{x \cdot E}{Z} \cdot sin(\theta) \tag{3-8}$$

$$Q = \frac{x \cdot E - (x^2 + y^2)}{Z} \cdot sin(\theta) + \frac{y \cdot E}{Z} \cdot cos(\theta) \tag{3-9}$$

With: x, y Operation point in complex voltage reference frame

P, Q Operation point in complex power reference frame

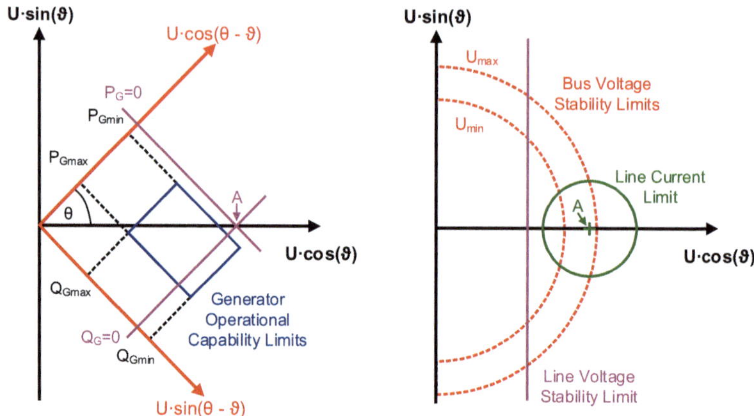

Figure 3-3: *Operational and stability constraints of two-bus system in the complex voltage reference frame. (left) Generator operational limits, (right) Grid constraints. [10]*

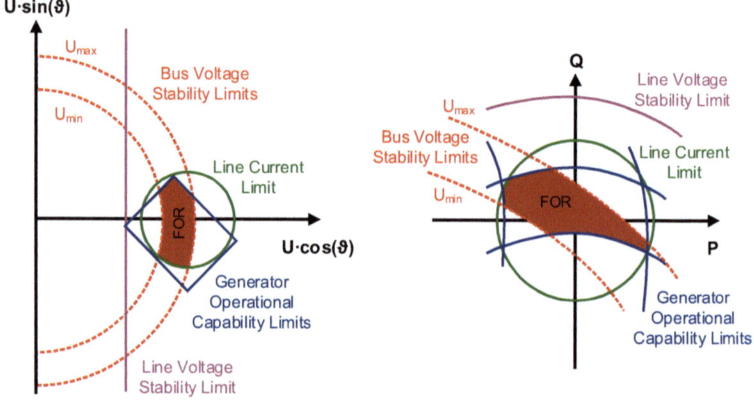

Figure 3-4: *Capability chart of synchronous generator considering grid and generator constraints. (left) Complex voltage reference frame, (right) Complex power reference frame. [10]*

In [109], the resulting capability chart for the two-bus system is contrasted with Newton-Raphson power flow (NR-PF) calculations for different operation points (P_G, Q_G) at the PQ bus. The valid operation points showed to be confined within the computed area, hence validating the model for this simple example. Based on [10], this analytical method could be extended to analyze the integration of a single generator in larger grids. This is performed by replacing the slack bus and the power line of Figure 3-2 with a Thévenin equivalent of the grid, which can be calculated using equations (3-10) to (3-12)[7]. The voltage in the Thévenin equivalent is computed according to the vectors of the higher and lower voltage solutions ($\underline{U}_H$ and $\underline{U}_L$) at the corresponding bus. Nevertheless, supplementary corrections are required to allow a proper representation of aggregated generator limits, as described in [10].

$$u_{TH} = \frac{\left|\underline{U}_H\right|^2 - \left|\underline{U}_L\right|^2}{\left|\underline{U}_H - \underline{U}_L\right|}$$
(3-10)

$$Z_{TH} = \frac{\left|\underline{U}_H\right|\left|\underline{U}_L\right|}{\sqrt{P_G^2 + Q_G^2}}$$
(3-11)

$$\theta_{TH} = \tan^{-1}\left(\frac{Q_G}{P_G}\right) - \vartheta_H - \vartheta_L$$
(3-12)

With: U_{TH} Magnitude of Thévenin equivalent voltage

         $\underline{Z}_{TH}, Z_{TH}\angle\theta_{TH}$ Complex Thévenin equivalent impedance

         $\underline{U}_H, U_H\angle\vartheta_H$ Highest complex voltage at PQ bus

         $\underline{U}_L, U_L\angle\vartheta_L$ Lowest complex voltage at PQ bus

Analytic methods are not the best approach to analyze the impact of several FPU scattered over a larger grid, as the grid equation systems can become extremely complex when the number of buses increases. This motivates the development of other types of methods to assess the FOR of power systems.

3.2.2 Geometric Approaches

The FOR of a power system results from the addition of the single capability charts of the FPU. These polygons are represented as a set of vectors in the $\Re^2$ space, where each vector is a vertex of the polygon. The Minkowski sum[8] (or dilation operator), is a mathematical method used to perform the addition of two or more polytopes [111]. The mathematical formulation for the addition of two polytopes P_1 and P_2 is provided in (3-13). The operator performs the sum of all vectors describing the vertices of both polytopes. A key property for its application in the computation of the FOR is that the convex hull operation applied to the Minkowski sums of two polytopes is a commuting operation, as described in (3-14) [111] [112]. This means that the order of the polytopes does not affect the result. A

---

[7] The norm of a complex number $\underline{X} = a + jb$ is defined as $\left|\underline{X}\right| = \sqrt{a^2 + b^2}$

[8] Minkowski sums can be applied to both convex and non-convex polytopes, however, the complexity of the algorithm strongly increases with non-convex polytopes.

simple example of the addition of two polytopes using the Minkowski sums is shown in Figure 3-5, including the commuting property of the operator.

$$P_1 \oplus P_2 = \{p_1 + p_2 | p_1 \in P_1 \wedge p_2 \in P_2\} \tag{3-13}$$

$$conv(P_1 \oplus \cdots \oplus P_n) = conv(P_1) \oplus \cdots \oplus conv(P_n) \tag{3-14}$$

With: P_k Polytope in $\mathfrak{R}^2$

        $conv(P)$ Convex hull operator applied to polytope P

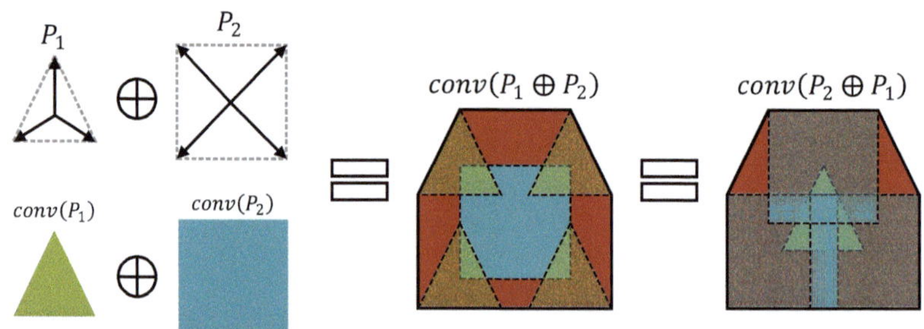

Figure 3-5: *Example of the Minkowski sum of two convex polytopes. Due to the commuting property of the convex hull operator, the resulting polytopes are identical. [112]*

In [97], [113], [114], [115] and [116], Minkowski sums are applied to the FOR computation. The complexity of the Minkowski increases with the number of FPU and the number of vertices of each capability chart, as the possible combinations of vectors increases exponentially. Therefore, in [97], a method to decrease the complexity was proposed, in which the polytopes are approximated to polygonal shapes with reduced number of sides (e.g. squares or hexagons). The approach makes good use of the general properties of polygons to perform the calculations, simplifying its implementation. Additionally, it provides an alternative to include discrete FPU models, i.e. switching a thermal load on and off, into the computation of the FOR. The main disadvantage is that the impact of the grid impedance and the grid constraints in general cannot be integrated into these methods.

3.2.3 Random Sampling Approaches

A method to compute the FOR of radial distribution grids based on Monte-Carlo simulations was proposed in [21]. In the algorithm, random operation points within the capability chart of each FPU are obtained and included into a load flow calculation, from which the IPF of the grid is computed. Each IPF is validated to see if the grid state that defines it is valid; all grid constraints need to be respected. If the IPF violates at least one constraint, it is marked as non-valid, otherwise it is defined as valid. After evaluating the validity of all randomly obtained IPFs, the

FOR is obtained by applying the convex hull operator to all valid IPF. This random sampling (RS) approach has been subject of plenty of research in the last years.

Modern load flow algorithms, together with modern computers, makes it possible to perform large amounts of load flow calculations in reasonable time. This makes the adoption of such RS methods to compute the FOR of interest, especially due to their relatively simple implementation. The following subchapters provides the mathematical description of the method in connection to convex FPU capability charts, for its application to radial distribution grids. Then, it is demonstrated how different probability distribution functions (PDF) for the selection of random FPU operation points can have a large impact in the shape of the assessed FOR.

3.2.3.1   Load Flow Calculation Considering FPU

The RS approach described in this subchapter is an adaptation of the model proposed in [21]. The complex power $\underline{s_i} = p_i + j \cdot q_i$ injected at each bus $i \in B$ (where B contains all buses in a grid) can be expressed in polar coordinates as in (3-15) and (3-16). These equations depend on the complex nodal voltages $\underline{u_i} = u_i \cdot e^{j \cdot \vartheta_i}$ and the complex branch admittances $\underline{y_{ij}} = y_{ij} \cdot e^{j \cdot \theta_{ij}}$ (from which the admittance matrix Y is built). The nodal power balance in normal steady-state operation is described by the non-linear equations (3-17) and (3-18), which shows the difference between the expected and the calculated (from (3-15) and (3-16)) complex power at bus i [117].

In [21], the complex power of an FPU is explicitly added to the power balance equations as a random variable between 0 and 1. The model described in this subchapter differs from the one in [21], as the active and reactive power operation points ($\{p_{FPU,f}, q_{FPU,f}\} \in \Re$) of a given FPU f are considered as independent random variables and added directly to the nodal balance equations, and not through an additional variable. Consequently, the complex power at any bus i is divided into a flexible and a non-flexible share, depending on what is connected to that bus. The non-flexible share represents the aggregated operation points of all non-flexible resources, while the flexible share includes all the FPU. In contrast to [21], which limits the FPU operation to just rectangular capability charts, the model described here allows using any type of convex capability chart for the FPU, as was described in the previous chapter.

The load flow problem seeks for a solution of the non-linear power flow equations, meaning the complex bus voltages $\tilde{u}_i$ that solve $\Delta p_i \approx 0 \wedge \Delta q_i \approx 0, \forall i$. Applying $\tilde{u}_i$ to (3-19) and (3-20) permits the computation of the branch flows through all branches in the grid. The IPF, which is the power flowing through the slack bus of the grid, is included in those equations. The process is repeated a predefined number of times (generally in the thousands or millions of repetitions), each time using new randomly selected operation points. After each load flow computation is performed, it is validated if the branch flows are located within the boundaries set by (3-21) and if the magnitude of the voltages $\tilde{u}_i$ are located within the limits defined in (3-22). An IPF is tagged as valid if each one of the grid constraints is respected, otherwise it is tagged as non-valid. In the end, a cloud of valid and non-valid IPF is obtained in the complex power domain. The FOR can be defined as a polygon that encloses a given set of valid IPF. Though, in general, the FOR is defined as the convex hull of all valid IPFs.

$$p_i(u, \vartheta) = u_i \cdot \sum_{j=1}^{n} y_{ij} \cdot u_j \cdot cos(\vartheta_{ij} - \theta_{ij}) \tag{3-15}$$

$$q_i(u, \vartheta) = u_i \cdot \sum_{j=1}^{n} y_{ij} \cdot u_j \cdot sin(\vartheta_{ij} - \theta_{ij}) \tag{3-16}$$

$$\Delta p_i = p_{fix,i} + \sum_{f \in F_i} p_{FPU,f} - p_i(u, \vartheta) \tag{3-17}$$

$$\Delta q_i = q_{fix,i} + \sum_{f \in F_i} q_{FPU,f} - q_i(u, \vartheta) \tag{3-18}$$

$$p_{ij} = u_i \cdot y_{ij} \cdot \left(u_i \cdot cos(\theta_{ij}) - u_j \cdot cos(\vartheta_{ij} - \theta_{ij})\right) \tag{3-19}$$

$$q_{ij} = u_i \cdot y_{ij} \cdot \left(u_i \cdot sin(-\theta_{ij}) - u_j \cdot sin(\vartheta_{ij} - \theta_{ij})\right) \tag{3-20}$$

$$p_{ij}^2 + q_{ij}^2 \leq s_{ij,max}^2 \tag{3-21}$$

$$u_{i,min} \leq u_i \leq u_{i,max} \tag{3-22}$$

With:

p_i, q_i		Calculated active/reactive power injection at bus i
p_{ij}, q_{ij}		Calculated active/reactive power flow from bus i to bus j
$\Delta p_i, \Delta q_i$		Mismatch between specified and calculated active/reactive power injections at bus i
$y \angle \theta_{ij}$		Complex admittance of branch connecting buses i and j
u_i		Voltage magnitude at bus i
$\vartheta_{ij} = \vartheta_i - \vartheta_j$		Voltage angle difference between buses i and j
$p_{fix,i}, q_{fix,i}$		Sum of non-flexible active/reactive power consumption at bus i
F_i		Set of FPU connected to bus i
$p_{FPU,f}, q_{FPU,f}$		Flexible active/reactive power consumption of FPU f in F_i at bus i
$u_{i,min}, u_{i,max}$		Min. and max. voltage magnitude at bus i
$s_{ij,max}$		Max. apparent power of branch between buses i and j

3.2.3.2 2D-Distributions for Random Sampling

The capability chart of an FPU f is generally simplified as a rectangle in the complex power domain, bounded by $p_{min} \leq p_{FPU,f} \leq p_{max}$ and $q_{min} \leq q_{FPU,f} \leq q_{max}$. Within these boundaries, random values for $p_{FPU,f}$ and $q_{FPU,f}$ are selected using different PDF. Different types of two-dimensional PDF can be found in literature related to the FOR computation. A Uniform PDF has been used in [7], [14], or [21], while [118] proposes a Normal PDF, and [119] makes use of a Beta PDF. What can be noticed in these cases is that the PDF used for the selection of the random FPU operation points has a noticeable impact in the resulting FOR, as the IPF strongly depends from the combinations of FPU operation points.

To assess the real boundaries of the FOR, it is necessary to explore combinations of operation points located at the edges of the FPU capability charts. These combinations are more likely to result in extreme FOR points, as was observed in [119]. In a most extreme case, combinations of only the vertices of rectangles can be obtained by applying a Rademacher PDF[9], which resembles a Beta PDF[10] with $\alpha, \beta \ll 1$. Using the Rademacher PDF resembles the use of the Minkowski sums, as shown in Chapter 3.2.2. As $p_{FPU,f}$ and $q_{FPU,f}$ are independent random variable, a bivariate heat map can be obtained from applying the described PDFs. The two-dimensional Normal, Uniform, Beta and Rademacher are shown in Figure 3-6. In the case of the Uniform PDF, all points within the boundaries have the same probability to be selected, hence the almost monochrome heat map.

It was shown in [119] that using a Beta PDF can help improving the assessment of the FOR compared to the Uniform PDF. In [108], this analysis was extended to the Normal and Rademacher PDF as well. The main findings in both papers is that the quality of the assessed FOR can be greatly improved by considering a PDF which focuses on the corners of rectangular FPU limits instead of the drawing operation points located in the center of the capability chart. All described PDF can be expressed with analytic equations, which eases their practical implementation; however, such continuous PDF are difficult to obtain for other FPU shapes different than a rectangle, especially when it comes to irregular convex polygons. This requires the development of alternative methods [108]. As was the case for the rectangular boundaries, the proposed PDF also focus on providing random operation points located closer to the vertices of the polygons.

---

[9] The Rademacher PDF is a discrete probability distribution where a random variate x has a 50% chance of being +1 and a 50% chance of being -1.

[10] The Beta PDF is defined as: $f(x; \alpha, \beta) = \dfrac{x^{\alpha-1} \cdot (1-x)^{\beta-1}}{\int_0^1 u^{\alpha-1} \cdot (1-u)^{\beta-1}\, du}$

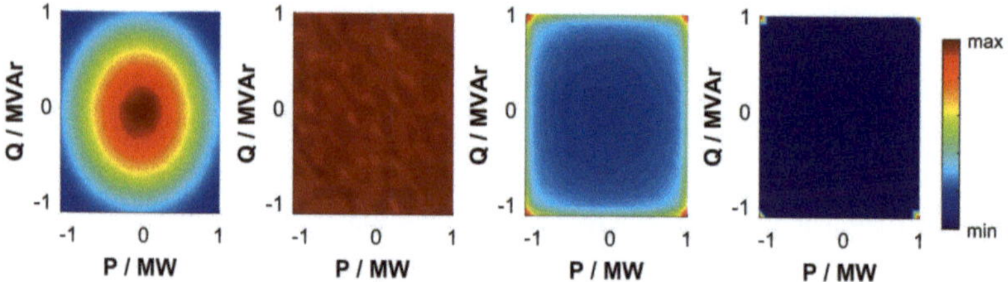

Figure 3-6: Heat maps of four different rectangular bivariate PDF. (left to right) Normal, Uniform, Beta, and Rademacher PDF. [108]

Two novel approaches were proposed in [108] with the objective of replicating the bivariate Beta and Rademacher PDF and allow their application to generic convex polygons. In the first approach, a discrete two-dimensional PDF is constructed based on the histogram of a discretized image containing a polygon. Each bin (i.e. pixel) containing the polygon is given a weight larger than zero, while the bins outside of the polygon always have a weight equal to zero. The weights of the bins are obtained by reconstructing the polygon using circles. Similar to the Beta PDF, the vertices of the polygon have the highest probability density, while the probability of returning a point in the center of the polygon is almost neglectable (but not zero). The algorithm is based on the proposal of [120] and the Medial Axis concept depicted in [121]. An example of the usage of the histogram-based PDF is shown in Figure 3-7.

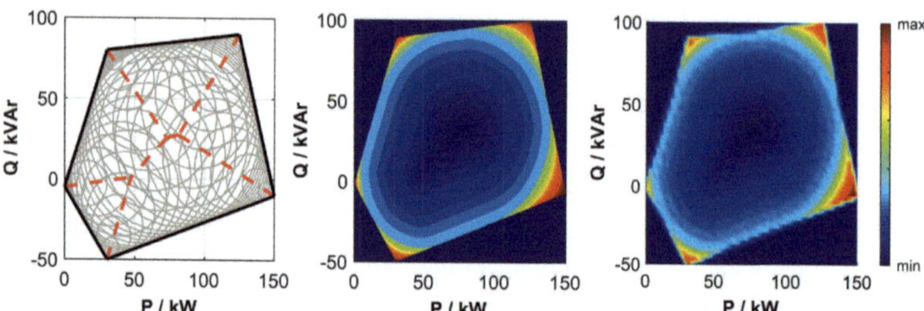

Figure 3-7: Generation process of bivariate PDF for irregular polygons from a histogram. (left) Medial axis and inscribed circles, (center) Constructed PDF from weighted topo-logical map, (right) Resulting heat map of random samples. [108]

The second approach mimics the Rademacher PDF, allowing its use in polygons with more than four vertices. Applying the RS method using operation points of each FPU located only at the vertices of their capability chart would allow for in-creasing the size of the assessed FOR, as a larger number of extreme load/gen-eration scenarios can be assessed. In [108], two variations of this approach were presented, which are described as follows:

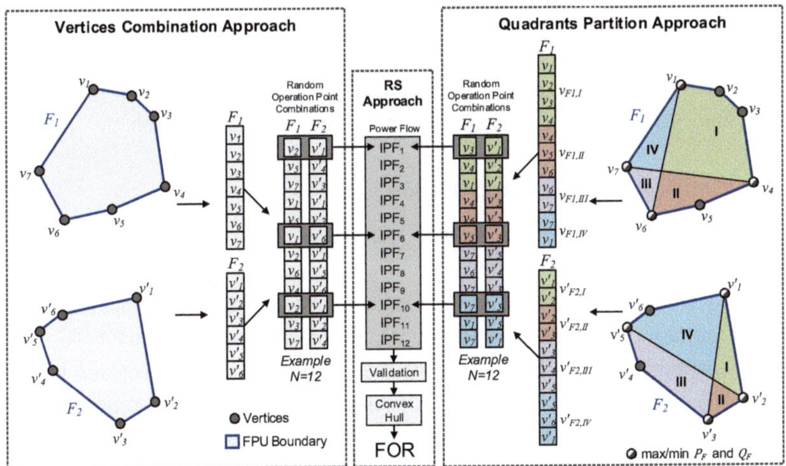

Figure 3-8: Illustration of the two proposed approaches for the selection of random samples considering only combinations of vertices; vertices reduction and quadrants partition [108].

- **Vertices Combinations:** This method is based on the reduction of the capability chart of an FPU (denoted as $\mathcal{F}$) to contain only the vertices $v_{\mathcal{F}}$ of the convex polygon (assuming $\mathcal{F}$ has at least two vertices). All other FPU operation points are neglected, both the interior ones, as well as the ones located at the edges of $\mathcal{F}$. The algorithm selects one random vertex from $v_{\mathcal{F}}$ of each FPU $\mathcal{F}$ in the grid uniformly. The process is repeated N times, resulting in N vectors, each with one random sample from each $\mathcal{F}$. The process resembles the Rademacher PDF, yet allowing for more than just two discrete states. The usage of the algorithm is illustrated in Figure 3-8 (left half). Some operation point combinations defined in $v_{\mathcal{F}}$ would not lead to maximal or minimal load/generation scenarios (e.g. $\min P_{gen}/\max P_{load}$ or $\max P_{gen}/\min P_{load}$), hence the approach would generate IPFs placed in the interior of the FOR, instead of expanding its area.

- **Quadrants Partition:** The method focuses on matching operation points that aim into the same direction in the complex power space. This is achieved by dividing the vertices $v_{\mathcal{F}}$ into four quadrants (I to IV) as shown in Figure 3-8 (right half). The quadrants are limited according to the maximal and minimal values of $P_{\mathcal{F}}$ and $Q_{\mathcal{F}}$, both inherent to the shape of $\mathcal{F}$. The four extreme points are automatically assigned to the two quadrants they divide. The result is a partition of the vector containing all vertex $v_{\mathcal{F}}$ into four different vectors $v_{\mathcal{F},I-IV}, \forall \mathcal{F}$. The algorithm proceeds to obtain random samples of vertices contained in each of the quadrants. From each quadrant, 25% of the total samples are obtained, resulting in N total samples.

The RS approach then combines random samples from the same quadrant for each FPU. This approach is similar to the objective functions of the OPF defined later in Chapter 4.2.9, where it is intended to expand the shape of the FOR as much as possible in one single direction at a time.

3.2.4 Optimization Based Approaches

An attempt to generalize the computation of the FOR of a power grid was made in [8], [9], and [105], with the formulation of an optimization problem to assess the capability of a power system. The formulation of the optimization problem resembles other traditional OPF approaches, only with a different objective function, as no operation costs were considered. As the grid constraints are included in the optimization problem, this would be a security constrained optimal power flow (SCOPF). However, as stated in [122], nowadays it makes no sense to formulate an OPF and not include the grid constraints, meaning that there is no need for the SCOPF concept anymore, as it is already contained in the OPF definition. Therefore, this work assumes the inclusion of grid constraints within the formulation of any OPF, including, but not limited to, voltage and branch loading limits.

In optimization-based approaches, solving an OPF results in a single vertex of the FOR. Therefore, an iterative process to assess the FOR was introduced in [8], where the active and reactive power participation factors of system loads are modified, changing the direction of search in the PQ cartesian plane. After repeating the process several times with different parameters, enough boundary points can be achieved and the FOR can be constructed.

A variation of the OPF problem was presented in [11], which permitted the inclusion of control techniques of DER into the problem, e.g. Volt/VAr control. Additionally, a transformer with line-drop compensation is considered. The objective function is adapted to optimize the reactive power exchange at the grid connection point, paying attention to the control strategies. The same optimization approach is applied in [12] to compute the reactive power provision potential at the interconnection point. An OPF approach is applied in [13] and [123] to compute the capability chart of a microgrid. The objective function controls the active power flow at the interconnection point, while the internal losses in the microgrid are minimized.

This chapter describes how the IPF located at the edge of the FOR can be assessed by solving an OPF problem. The OPF can be formulated according to the aforementioned proposals, however, the different objective functions can be summarized into a single global function. It is logical that the FOR cannot be assessed in a single step, therefore, an algorithm to iteratively explore the FOR boundary solving several OPF is at all times necessary. Many different concepts to optimize the usage of the OPF in order to assess the boundary points of the FOR have been proposed in [22], [103], [124], and [125], some of which are discussed in this chapter as well. As an alternative to classical OPF approaches, a stochastic optimization approach is proposed in [126] with the focus of controlling the power exchange at the TSO-DSO interface.

3.2.4.1   General formulation of an OPF

An OPF is an optimization method that considers the power flow equations. The goal is to find an optimal operating level of a power system that satisfies the security constraints (e.g. nodal voltages, branch loading limits, capability chart of synchronous generators or DER, cost functions, voltage controllers). The constraints are formulated as a set of equalities and inequalities, which can be linear or non-linear. The general formulation of an OPF is shown (3-23). In standard OPF problems, the state vector x contains the nodal voltages $\underline{u}_i$ of the grid, as well as other desired decision variables, conditioned to the overall goals of the OPF. For the assessment of the FOR, state vector x includes the operation points of the FPU f ($P_{FPU,f}$ and $Q_{FPU,f}$) as well, as their capability charts are what defines the shape of the resulting FOR. The operation of an OLTC transformer could be included in x in the form of a discrete tap position or a continuous transformation ratio (e.g. [22] and [124]). Yet, it was shown in [23] that including the tap position as a decision variable can produce a false FOR.

$$\min_{x} f(x) \; s.t. \begin{cases} c_{ineq}(x) \leq 0 \\ c_{eq}(x) = 0 \\ A_{ineq} \cdot x \leq b_{ineq} \\ A_{eq} \cdot x = b_{eq} \\ x_{lb} \leq x \leq x_{ub} \end{cases} \tag{3-23}$$

With:

$f(x)$	Objective function
x	State variables vector
$c_{ineq}(x)$	Non-linear inequalities
$c_{eq}(x)$	Non-linear equalities
A_{ineq}	Matrix with left-hand-side of linear inequalities
b_{ineq}	Vector with right-hand-side of linear inequalities
A_{eq}	Matrix with left-hand-side of linear equalities
b_{eq}	Vector with right-hand-side of linear equalities
x_{ub}, x_{lb}	Upper/Lower boundaries of vector x

The power flow problem was already defined in equations (3-15) to (3-20). To define the OPF, equations (3-17) and (3-18) need to be rewritten as equalities. As

$p_i(u, \vartheta)$ and $q_i(u, \vartheta)$ are non-linear functions, the optimization problem is non-linear as well.

$$\Delta q_i = p_{fix,i} + \sum_{f \in F_i} p_{FPU,f} - p_i(u, \vartheta) = 0 \qquad (3\text{-}24)$$

$$\Delta q_i = q_{fix,i} + \sum_{f \in F_i} q_{FPU,f} - q_i(u, \vartheta) = 0 \qquad (3\text{-}25)$$

Grid constraints are added as inequalities in the form of (3-21) and (3-22), while the FPU constraints are represented, initially in a simplified form, as rectangles in the complex power domain:

$$p_{FPU,f,min} \leq p_{FPU,f} \leq p_{FPU,f,max} \qquad (3\text{-}26)$$

$$q_{FPU,f,min} \leq q_{FPU,f} \leq q_{FPU,f,,max} \qquad (3\text{-}27)$$

In [22], [123], and [125], the objective of the OPF is to optimize the power flow at the interconnection point of the distribution grid with an overlayered grid. This power flow was defined as the IPF in the previous chapter. The following general formulation of the objective function is then provided:

$$f(x) = \alpha \cdot p_{IPF}(x) + \beta \cdot q_{IPF}(x) \qquad (3\text{-}28)$$

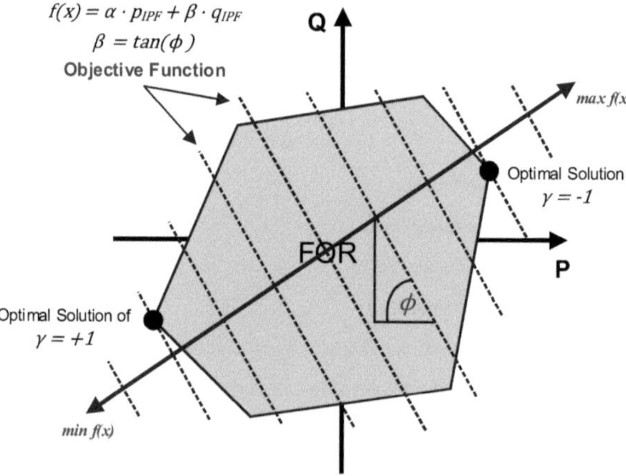

Figure 3-9: Usage of the OPF objective function with values of parameters α and β and the assessment of boundary points as minimization and maximization problems.

The IPF is contained in (3-19) and (3-20), when considering the single branch connecting the slack bus to the grid (e.g. a HV/MV transformer) or the sum of all branches connecting to the slack bus (e.g. LV feeders). Both parameters α and β allow controlling the direction in which a boundary point of the FOR is searched. Figure 3-9 provides a schematic representation of the objective function in relationship to the FOR. For each combination of α and β, the search direction (blue arrow in Figure 3-9) would rotate, meaning that a proper parametrization would allow exploring the FOR in all 360°. In optimal scenarios, each new OPF is solved with a different parametrization of the objective function. This requires developing techniques which optimize the usage of the OPF when assessing the FOR.

3.2.4.2 Review of OPF-based Algorithms to Assess the FOR

Numerous OPF-based algorithms for the assessment of the FOR of radial distribution grids have been proposed in the last years. Some variations of OPF-based approaches are introduced in [9], [22], [55], [123], [124], and [125]. The common goal is to solve a non-linear OPF. At the same time, these methods provide several techniques to apply the OPF to assess the entirety of the FOR. As the real shape of the FOR is initially unknown, iterative processes are required to reconstitute its shape in the best possible way, which results in the approximation of the FOR through a convex irregular polygon. It is assumed that a convex solution space will be obtained, since both the FPU and the grid constraints are convex [127] [128]. This statement may become invalid if the FPU boundaries are deemed as voltage dependent [86] [129].

If the FOR is indeed assumed convex, considering aforesaid restrictions, the simplest approximation is given by locating the four extreme values of the perimeter using (3-28) with the combination of parameters shown in Table 3-1. By obtaining the four extreme IPF, a rough approximation of the FOR can already be achieved, as it can be represented through a rectangle, as shown in Figure 3-10 [22] [23].

Table 3-1: Parametrization of equation (3-28) to obtain four extremes FOR boundary points[11].

	Parameters	
IPF	α	β
P_{max}	-1	0
P_{min}	1	0
Q_{max}	0	1
Q_{min}	0	-1

---

[11] In this chapter, the notation P and Q describes FOR boundary points in the complex power domain. For example, P_{min} describes the boundary point defined by the minimal active power value, while Q_{min} describes a point with the minimal reactive power value.

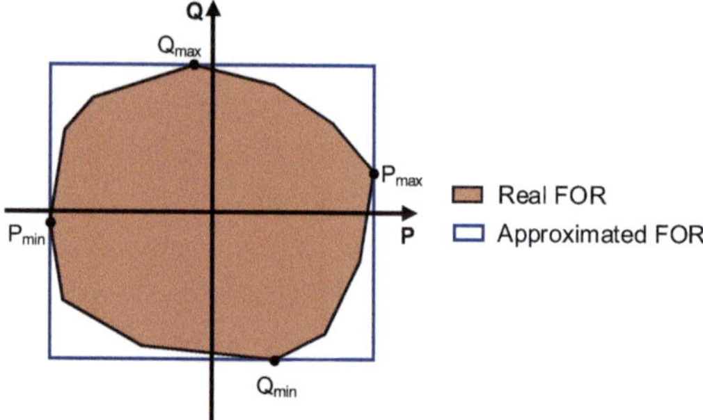

Figure 3-10: Rough approximation of the FOR by a rectangle defined by the four extreme IPF.

The representation of the FOR through a rectangle provides a simple and practical solution to the problem at hand, yet it usually overestimates the valid operation region. This can be improved by increasing the amount of FOR boundary points explored. Both [11] and [55], propose optimizing the reactive power flow q_{IPF} using constant values of the active power flow p_{IPF}, as illustrated in Figure 3-11. The values of p_{IPF} are in this case obtained from time-series of the DER active power injection. A similar proposal is found in [123], where the objective function focuses on the optimization of p_{IPF} for constant reactive power transfers q_{IPF}.

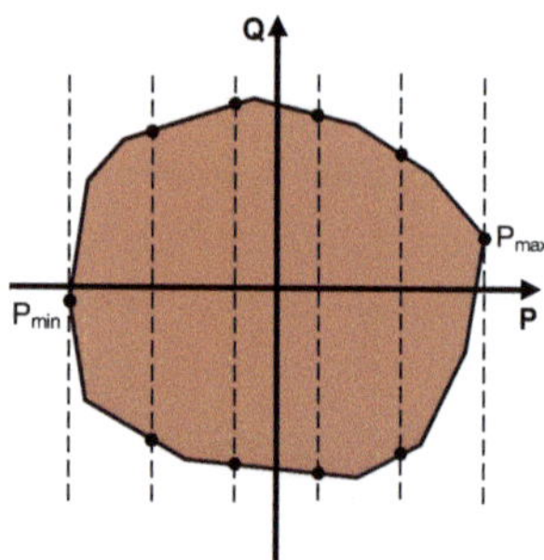

Figure 3-11: FOR assessment with constant active power flow p_{IPF}. [11] [55].

In [124] and [125], both the active power and reactive power flows (p_{IPF} and q_{IPF}) are optimized sequentially. Based on the extreme values, P_{min} and P_{max}, a lattice-based search is performed with discrete increments of p_{IPF}, and for each value the OPF is solved. This way the boundary points Q_{min} and Q_{max} are obtained. Then, the process is repeated, only this time the OPF is solved with constant reactive power values, as shown in Figure 3-12. A smaller discretization of the steps allows to improve the quality of the assessed FOR, at the cost of performing additional OPF, which increases the running time of the algorithm.

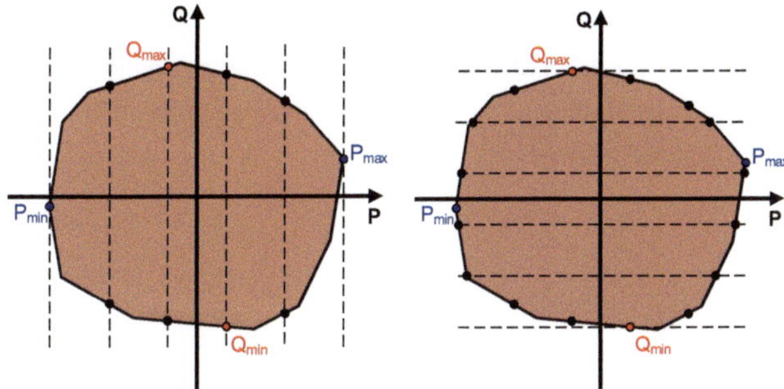

Figure 3-12: Lattice-based FOR assessment approach. (left) Optimization with constant active power exchange, (right) Optimization with constant reactive power. [124] [125]

A different approach was proposed in [9], [103], and [127], where the active power exchange p_{IPF} is optimized with a constant power factor ϕ ($q_{IPF} = p_{IPF} \cdot \tan(\phi)$), therefore, placing a linear dependence between p_{IPF} and q_{IPF}. This results in a search for boundary points in a straight line in different directions from a specific initial operation point, as depicted in Figure 3-13. Generally, the origin is used as the initial operation point, from which the search begins. However, there are some situation in which this method would not work properly, especially when the FOR does not include the origin of the complex power domain.

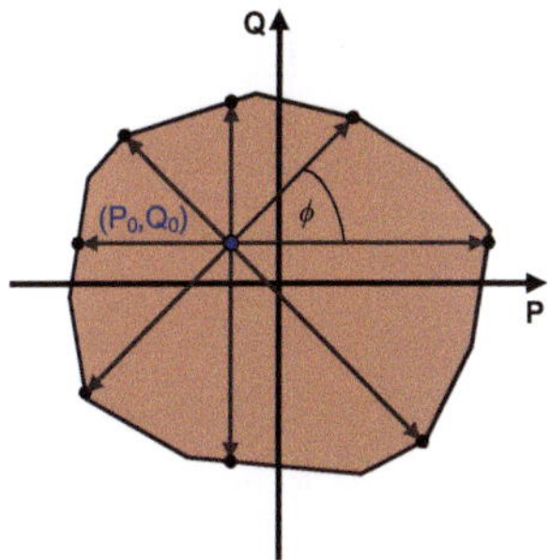

Figure 3-13: FOR assessment with constant power factor. [9] [103]

In [22] and [23], two approaches were proposed which combine the aforementioned lattice-based and constant power factor methods, making full use of (3-28). In the approach of [22], the first step involves assessing 8 FOR boundary points are with constant power factor values defined by $\phi = \left\{0, \frac{\pi}{8}, \frac{2\pi}{8}, \frac{3\pi}{8}, \cdots, \frac{7\cdot\pi}{8}\right\}$. The extreme values Q_{min} and Q_{max} are obtained in this step. In the second step, an iterative process takes place, in which the OPF is solved with constant reactive power values. In this iterative process, a new constant value for q_{IPF} is obtained by averaging the reactive power value of two already assessed adjacent boundary points $(\frac{1}{2}(Q_i + Q_{i+1}))$. The process is illustrated in Figure 3-14. The results are refined iteratively, and each iteration adds new boundary points, which are also considered in the computation of the next iterations. The process is repeated, until the Euclidean distance between all neighbors is small enough.

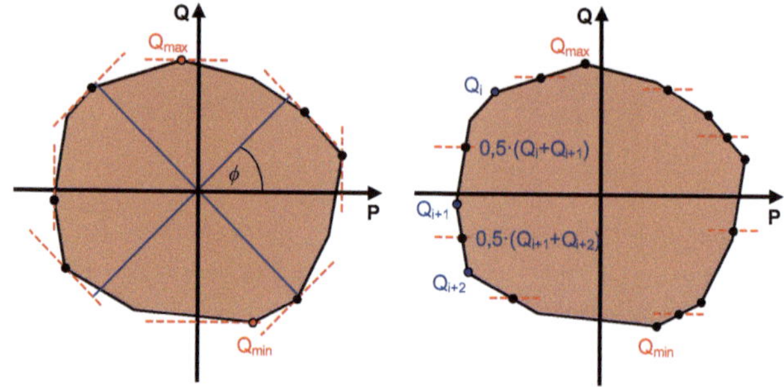

Figure 3-14: Combined FOR exploration approach. (left) Optimization with constant power factor, (right) Optimization with constant reactive power. [22].

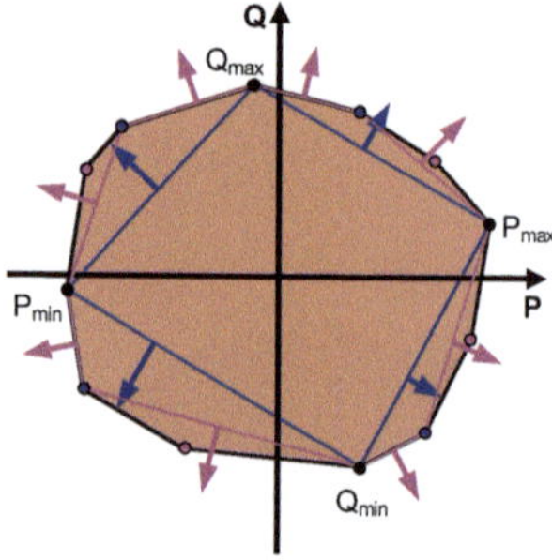

Figure 3-15: Iterative FOR exploration approach based on self-adaptive tangent lines. [23]

The ICPF algorithm of [23], proposes to iteratively increase the size of the assessed FOR by directly optimizing the parameters α and β in (3-28). The objective function of the OPF is defined as the tangent of the contour of the FOR. This requires an initial assessment of four extreme boundary points based on Table 3-1 and connecting these points with straight lines (blue lines in Figure 3-15). Each line represents a tangent of the FOR, which is defined through the four initial points, and they are modelled based on (3-28), for which specific values for α and β are computed. The OPFs are solved again with the new objective functions and these are used to obtain a new set of boundary points (blue markers in Figure 3-15), by performing the optimization in the orthogonal direction of the tangents. The process is repeated iteratively including all the newly assessed (magenta lines and markers in Figure 3-15), until convergence is reached. Two possible stop criteria have been proposed in literature, on one side, the Euclidean distance from two adjacent points in the FOR, and on the other, the angle difference between two adjacent internal corners of the FOR.

All the aforementioned procedures have in common that they solve an OPF with (3-28) as objective functions. The difference between them lies only in the parametrization of α and β. In the end, each algorithm results in a piecewise linear approximation of the FOR, however, for the same scenario, the different search procedures can lead to different FOR. The following is a summary of the main aspects of the presented methods:

- **Number of Iterations**: In [11], [55], [124], and [125], the granularity of the mesh in the lattice-based algorithms needs to have to be defined in advance, as the methods are not self-adaptive. A finer mesh would lead to an increase in the number necessary OPF that need to be solved, in most cases leading to many avoidable iterations which do not provide additional information that allow increasing the size of the FOR. This issue was tackled in the iterative approaches of [22] and [23], which optimized the number of necessary OPF, as they constantly adapt the parametrization of (3-28) to the shape of the FOR. In contrast to lattice-based methods, where the number of iterations is fix, the iterative ones require explicit break criteria, which need to be carefully selected.

- **Resulting Shape of FOR**: The shape of the FOR is defined by all the defined convex constraints. Linear constraints will result in a FOR with sharp edges. The convexity of the FOR depends directly on the type of constraints in the OPF, and the modelling of the power flow equations. If the FOR has vertices with straight or acute internal angles, this poses additional challenges to iterative approaches, as they may reach repeatedly to the same IPF, requiring additional procedures to break the algorithm in case of non-convergence [23].

- **Impact of Initial Operation Point**: Some methods are sensitive to the initial operation point of the grid and to the location of the FOR in the cartesian plane (e.g. [9] and [103]). Generally, the initial operation point should be located within the FOR. Cases where the FOR does not contain the origin of the complex power domain should be further studied.

- **Computational Time**: The outcome of [23] was a significant time reduction in the FOR calculation compared to the Monte-Carlo method in [21]. However, this was the only publication referring to the computational time issue. In general, OPF methods are faster than Monte-Carlo methods, however, the type of OPF (e.g. linear or non-linear) and the number of iterations, are the key factors that define the computational time of OPF methods.

4 Fast Computation of FOR with Linear Optimization

4.1 Method Objectives and Differentiation

A comprehensive summary of methods that allow assessing the FOR of radial distribution grids were introduced in the previous chapter. Most of them have been published just in the recent decade. A shared concern among these publications has to do with the computation time required to obtain the FOR, which in some cases can be excessive, i.e. [7], [23], [24] and [130]. Whereas long computation times may not be critical for long-term power system planning processes, it makes these methods unsuitable for the operation of power systems, where time-critical decisions need to be made and extremely complex and time-consuming methods are not helpful.

This acted as the main motivation for the development of a novel FOR assessment method throughout this thesis, the Linear Flexibility Aggregation (LFA) method, with the following goals:

- Minimize required computation time

- Assess radial distribution grids with large numbers of buses and FPU

- Include grid constraints

- Consider non-rectangular capability charts of FPU

The main innovations of the LFA method proposed in this work are the formulation of an entirely linear OPF problem to solve the FOR problem, including a linear power flow model and a set of linear constraints of the grid and the FPU. On one side, solving a linear OPF offers a remarkable reduction in the computation time, compared to non-linear optimization problems. On the other, according to [127] and [128], an OPF defined by a set of convex linear equalities and inequalities would lead to a convex solution space, critical for the overall optimization of the method. Additionally, this new method optimizes the search procedure of boundary points of the FOR and bounds the number of OPF that need to be solved, hence, allowing for an overall reduction of the computation time.

The way in which the LFA approach is defined allows for the assessment of larger radial distribution grids, with large numbers of FPU. This provides for a solution to an aspect that strongly affects the quality of the FOR provided by random sampling methods and greatly increases the complexity and running time of non-linear OPF methods, as is shown later in Chapter 5. Consequently, the LFA method provides a reliable and efficient tool to compute the FOR. The reduction in the computation time allows for the development brand new tools and decision-making mechanisms for the planning and operation of power systems. Some of which are described later in Chapter 6.

This chapter provides a comprehensive explanation of the LFA method, including the description of the linear OPF model in 4.2. Then, 4.3 presents linear capability charts for FPU, based on the diagrams in Chapter 2.4. The validity of the proposed method and the impact of the grid topology in the results is discussed in 4.4. This chapter closes in 4.5 with an approach to couple the FOR computation with time-series of power injection and consumption.

4.2   Linear OPF-based Method to Compute the FOR

An OPF is a non-linear and non-convex problem by nature, due to the definition of the power flow equations. This type of problem is classified as NP-hard, meaning that the solution is not trivial and that a result cannot always be obtained in polynomial time. Yet, the most critical aspect is that a global optimum cannot always be ensured [122] [131] [132]. Such issues have been studied for decades and different solutions have been provided, each one suitable for a specific use of the OPF. The first concern that needed to be solved, is to ensure the convergence of the optimization, which usually comes at the cost of reducing the precision of the result. This can be performed by providing a convex relaxation of the constraints and/or by linearizing the grid model [132]. A comprehensive explanation of the impact of a convex relaxation in OPF problems is provided in [131].

This has given place to many different alternative proposals of OPF computation methods. One example, is the DCOPF, which approximates the power flow equations by neglecting reactive power flow [117], commonly used for market-based analysis in transmission grids (assuming $X \gg R$). In LV grids, the *DistFlow* approach is commonly used, which proposes a decoupled calculation of active and reactive power in an iterative way, although only applicable to radial grids [133]. Both these approaches to simplify the power flow problem cannot be properly applied to distribution grids, where $X/R \approx 1$, because most of the simplifications do not hold anymore.

There are several methods to solve non-linear power flow equations systems. Many of them, including the traditional Newton-Raphson method, postulate iterative solutions of the linearized equation system. The same principle can be applied to the development of the OPF necessary to compute the FOR. The linearization of the power flow equations follows two different objectives simultaneously. On one side, it helps ensuring the convergence of the OPF problem [127] [128], while on the other, it helps reducing the computational complexity, as solving a linear optimization problem can be much simpler than solving a non-linear one. This chapter describes the construction of a fully linear OPF approach that can be used to compute the FOR.

4.2.1 General Formulation of a Linear OPF

The purpose of a linear optimization problem is to calculate a vector x that minimizes the linear objective function $f(x)$ described in (4-1). The solution needs to satisfy a set of linear restrictions in the form of equalities and inequalities formulated according to (4-2). If the restrictions are convex, the feasible region containing all solutions of the OPF should be confined within a convex n-polytope. If the objective function is convex as well, the local minimum will be equal to the global minimum, meaning that the solution is unique [128]. The linear OPF can be defined similarly to the non-linear OPF described in (3-23).

$$f(x) = c^T \cdot x = c_1 \cdot x_1 + \cdots + c_n \cdot x_n \tag{4-1}$$

$$\min f(x) \; s.t. \; \begin{cases} A_{eq} \cdot x = b_{eq} \\ A_{ineq} \cdot x \leq b_{ineq} \\ x_{lb} \leq x \leq x_{ub} \end{cases} \tag{4-2}$$

With: $f(x)$ Linear objective function

         x Vertical state variables vector

         c^T Horizontal vector with linear equation coefficients

The power flows within the grid are described as nonlinear equations, which makes solving the OPF more difficult. It is known that solving a linear optimization problem, involves much less complexity that solving a non-linear one, while guaranteeing that a global optimum can be reached. This statement is valid under the premise that the convexity of the solution space can be ensured. The formulation of the different components of the OPF formulation of (4-1) and (4-2) are described in the upcoming chapters.

4.2.2 Linearization of Power Flow Equations

The standard power flow equations system in polar coordinates was described in Chapter 3.2.3.1. The real and imaginary parts (active and reactive power) of the complex power $\underline{s}_i$ injected at each bus $i \in B$ are described in (3-15) and (3-16). The non-linear equations (3-24) and (3-25) describe the nodal power balance achieved in steady-state operation [117]. Power flow algorithms have the task to find the complex voltages $\underline{u}_i, \forall i \in B$ that provide an (almost) exact solution to (3-24) and (3-25). This means that the nodal power injections p_i and q_i (from (3-15) and (3-16)) need to match the expected values that are defined for each node (including non-flexible and flexible power). It is not trivial to find an exact solution to the problem; therefore, an iterative approach to approximate (3-24) and (3-25) to zero ($\Delta P_i \approx 0 \wedge \Delta Q_i \approx 0, \forall i \in B$) becomes necessary. This formulation assumes all buses as PQ (minus the slack-bus, which has to be unique).

The Newton-Raphson (NR) method is an iterative algorithm that allows finding the root of nonlinear functions ($F(x) = 0$) efficiently. It is based on the notion that any differentiable function can be approximated at any point by a tangential line to the function and it is trivial to find the zero-crossing of a straight-line function. This results in the iterative process depicted in Figure 4-1 for a generic function $F(x)$. The kth iteration of the single variable NR method can be described as (4-3). During each iteration, x is corrected based on the slope of the derivative of $F(x)$. The process is repeated until the difference between two iterations becomes small enough ($|\Delta x| < \varepsilon$, with ε as a given tolerance).

$$x_{k+1} = x_k + \left(\frac{dF(x_k)}{dx}\right)^{-1} \cdot F(x_k) \tag{4-3}$$

$$x_{k+1} = x_k + \Delta x_k = x_k + J_k^{-1} \cdot F(x_k) \tag{4-4}$$

With: $F(x)$ Function to be linearized

         J, J^{-1} Jacobian of F and its inverse

         x_k Vector with state variables during kth iteration

         Δx_k Correction factor of state variables x during kth iteration

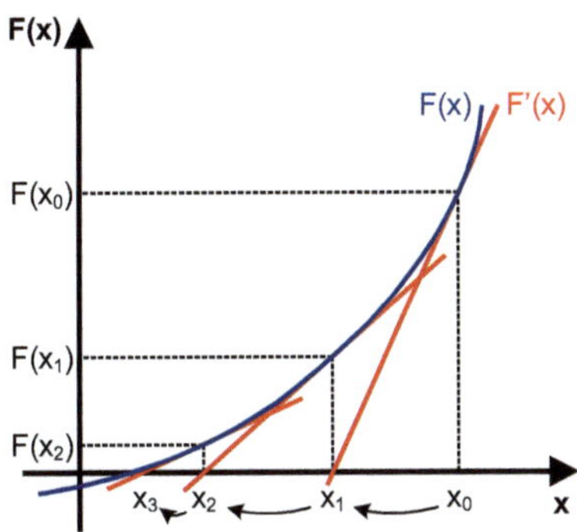

Figure 4-1: Exemplified procedure of finding the root of a one-dimensional function $F(x)$ using the iterative Newton-Raphson method.

In a multi-dimensional problem, i.e. the power flow problem, the inverse of the Jacobian matrix J of $F(x)$ is used to correct vector x, as shown in (4-4). Applied to the power problem, all components of J can be described analytically by deriving (3-15) and (3-16) with respect to ϑ and u. A full description of the equations is provided in Appendix D. During each iteration, the matrix J is updated based on the correctios to the state vector x, analogous to the one-dimensional approach.

The power flow problem searches for a combination of nodal voltages in the grid that complies with $|\Delta P_i| < \varepsilon$ and $|\Delta Q_i| < \varepsilon, \forall i$ (with ε an arbitrary small value). This is equivalent to finding the roots of (3-24) and (3-25). Applying the procedure described in (4-4) to solve this problem results in the well-known Newton-Raphson power flow algorithm. Through successive linearization of (3-24) and (3-25) using first order Taylor expansions, a linear equations system is obtained on each iteration. The function $F(x)$ in (4-4) is composed of (3-24) and (3-25), while the state vector x contains the voltage angles and magnitudes of all buses ($\vartheta_i, u_i, \forall i \in N$), including the slack bus. Accordingly, (4-4) is rewritten as (4-5), which displays the kth iteration of the NR-PF. The slack bus voltage remains constant and acts as a reference for the other voltages in the grid, therefore, $\Delta\vartheta_{slack} = \Delta u_{slack} = 0$. This results in the slack bus being removed from equation.

$$\begin{bmatrix}\Delta\vartheta_i \\ \Delta u_i\end{bmatrix}_{k+1} = \begin{bmatrix}\dfrac{\partial p_i}{\partial\vartheta} & \dfrac{\partial p_i}{\partial u} \\[2ex] \dfrac{\partial q_i}{\partial\vartheta} & \dfrac{\partial q_i}{\partial u}\end{bmatrix}_k^{-1} \cdot \begin{bmatrix}\Delta p_i \\ \Delta q_i\end{bmatrix}_k = J_k^{-1} \cdot \begin{bmatrix}\Delta p_i \\ \Delta q_i\end{bmatrix}_k, \forall i \in \{B - slack\} \qquad (4\text{-}5)$$

With: $\Delta\vartheta_{i,k}$ Difference of voltage angle for bus i on k[th] iteration

    $\Delta u_{i,k}$ Difference of voltage magnitude for bus i on k[th] iteration

    J_k^{-1} Inverse Jacobian Matrix J during the kth iteration

The state vector x is updated after each iteration, accordingly, (3-24) and (3-25) are updated after each iteration. As shown in (4-5), during each iteration, a linear equations system is solved. This is what allows the NR-PF algorithm to converge in just a few iterations in most cases. A critical convergence criterion is that J should be always invertible. For the first iteration of (4-5), an initial estimation of the voltages is required. Typically, a "flat-start" ($\underline{u}_i = 1\,p.u., \forall i \in B$) provides good results, as the grid is typically intended to be operated at voltages within the 0.9-1.1 p.u. range (e.g. [134]). Providing an initial guess even closer to the operation point of the grid may improve the convergence of the PF approach, although this is usually difficult to achieve as in reality grid observability is generally limited. The ability of the NR-PF to start from a "flat-start" is exploited in the proposed model to obtain an approximated operation point of the grid, around which the power flow equations are later linearized. According to [128], a linearized model can provide a unique solution for the power flow problem.

After $k = n$ iterations of (4-5), convergence is reached (when $|\Delta x| < \varepsilon$) and the resulting state vector $x_{k=n}$ is obtained, which in the LFA algorithm is defined as x_0, as it corresponds to the initial operation point of the proposed linear OPF model. When the inverse of J is evaluated in x_0, a linear power flow model is obtained. This results in $J_{x_0}^{-1}$, which is used as a linearization constant, resulting in the simplified power flow model depicted in (4-6). The linearization occurs around the grid operation point x_0, which is obtained from the previously described NR-PF computation based on a given grid operation point.

$$\begin{bmatrix} \Delta\vartheta_i \\ \Delta u_i \end{bmatrix}_x = \begin{bmatrix} \vartheta_i \\ u_i \end{bmatrix}_x - \begin{bmatrix} \vartheta_i \\ u_i \end{bmatrix}_{x_0} = J_{x_0}^{-1} \cdot \begin{bmatrix} \Delta p_i \\ \Delta q_i \end{bmatrix}_x , \forall i \in B \tag{4-6}$$

With: x_0 State variables vector containing initial grid operation point

        $J_{x_0}^{-1}$ Inverted Jacobian Matrix evaluated at initial state vector x_0

As the slack bus remains constant, then at all times $\Delta u_{slack} = 0$ and $\Delta\vartheta_{slack} = 0$. These expressions need to be added explicitly to (4-6), as for the calculation of the bus admittance matrix Y and the Jacobian J, the slack bus voltage is removed to ensure that J is invertible. Procedurally, this issue can be corrected by adding two additional rows and columns filled with zeros to $J_{x_0}^{-1}$, to ensure that $\Delta u_{slack} = 0$ and $\Delta\vartheta_{slack} = 0$.

Any change in the operation point of an FPU will cause Δp_i and Δq_i for some buses i to change, which then causes the values of ϑ and u within state vector x to change according to $J_{x_0}^{-1}$. Based on (4-6), the power flow can be easily computed through a single matrix multiplication, which drastically increases the computational speed compared to any iterative approach. On the downside, an error is expected to appear in the solution due to the linearization, which in the NR-PF algorithm is reduced on each iteration and in this case cannot be avoided. A more precise analysis of this error is performed in subsequent chapters.

According to equations (3-24), (3-25) and (4-5), J is directly coupled with the bus admittance matrix Y of the grid. This implies that any change in the grid topology, e.g. the operation of a switch or an OLTC transformer, will have an impact on Y and consequently on J and the inverse. As a result, the tap position of the OLTC transformer and the on/off state of a power line cannot be considered as independent decision variables in the proposed model, as was the case in [22] and [124]. For each topology change, both matrices Y and J are recalculated, generating a new linear model for each topology change and each operation point.

4.2.3 Adding FPU Flexibility into State Variables Vector

The flexible operation of the FPU can be added to the OPF formulation using the decision variables $p_{FPU,f}$ and $q_{FPU,f}$ according to (3-24) and (3-25). As not every load, generator or storage system connected the grid provides flexibility, the injected power at a bus is split between flexible and non-flexible terms. The non-flexible part (p_{fix}, q_{fix}) is the expected operation point of every non-flexible utility, while the flexible part $(p_{FPU,f}, q_{FPU,f})$ corresponds to the flexibility provided by the FPU $f \in F_i, \forall i \in \{B - slack\}$. In order to include the FPU capability charts into the OPF; the operation points of the FPU needs to be added to the state vector x.

The state vector x is shown in (4-7). It should be noted that x contains the slack bus voltage. This inclusion is essential in order to allow adding the loading constraint of the branch connecting the slack bus to the grid (i.e. a power line or, more likely, a transformer). Therefore, the slack bus voltage $u_{slack} \angle \vartheta_{slack}$ is required as an input to allow the initialization of the algorithm.

$$x = \begin{bmatrix} \vartheta_i \\ u_i \\ P_{FPU,f} \\ Q_{FPU,f} \end{bmatrix}, \forall i \in B, \forall f \in F_i \tag{4-7}$$

With: $P_{FPU,f}, Q_{FPU,f}$ Active/reactive power flexibility of FPU F_i at bus i

4.2.4 Linearization of Branch Flow Equations

The active and reactive power flows through the branch connecting buses i and j were already defined in (3-19) and (3-20). Both these equations are nonlinear because state variables u and ϑ are tightly coupled within them. In order to reach a fully linear OPF model, equations (3-19) and (3-20) need to be linearized as well. This also permits their inclusion in the branch flow constraints. The linearization process is similar than of the nodal equations. The Jacobian matrix of (3-19) and (3-20) is computed using the same linearization operation point x_0 as in (4-6). A detailed description of the Jacobi matrix is shown in Appendix D. The linear equations system for the branch flows is defined according to x as follows:

$$\begin{bmatrix} p_{ij} \\ q_{ij} \end{bmatrix}_x = \begin{bmatrix} p_{ij} \\ q_{ij} \end{bmatrix}_{x_0} + \begin{bmatrix} \Delta p_{ij} \\ \Delta q_{ij} \end{bmatrix}_x = \begin{bmatrix} p_{ij} \\ q_{ij} \end{bmatrix}_{x_0} + \begin{bmatrix} \dfrac{\partial p_{ij}}{\partial \vartheta} & \dfrac{\partial p_{ij}}{\partial u} \\ \dfrac{\partial q_{ij}}{\partial \vartheta} & \dfrac{\partial q_{ij}}{\partial u} \end{bmatrix}_{x_0} \cdot \begin{bmatrix} \Delta \vartheta_i \\ \Delta u_i \end{bmatrix}_x, \forall i,j \in B \tag{4-8}$$

With: $\Delta p_{ij}, \Delta q_{ij}$ Deviation of active/reactive power flow from bus i to bus j

4.2.5 Voltage and Branch Flow Constraints

Grid constraints are added to the OPF in two ways. First, the magnitudes of the nodal voltages u_i are constrained to a maximum and minimum boundary.

$$u_{i,min} \leq u_i \leq u_{i,max}, \forall i \in B \tag{4-9}$$

With: $u_{i,min}, u_{i,max}$ Voltage min/max boundaries

For the slack $u_{slack,min} = u_{slack,max}$, to ensures it remains constant, however, not necessarily at 1 p.u., as described later on in Chapter 5.

Second, the power flowing through the branches are restricted by the maximal apparent power of each branch (power lines or transformers), which are traditionally formulated through a quadratic equation:

$$\sqrt{p_{ij}^2 + q_{ij}^2} \leq s_{ij,max} \tag{4-10}$$

With: $s_{ij,max}$ Maximal apparent power of branch connecting buses i and j

Based on [117], [135] and [136], (4-10) can be piecewise linearized by approximating the boundary to an n-sides regular polygon. This does not compromise the numerical stability of the optimization problem. The apparent power limit of each branch $s_{ij,max}$, which results in a circle in the complex power domain, is then approximated piecewise by a total of m segments ($s'_{ij,max}$). Each segment is described through a straight-line equation $L_{ij,l}$ ($l = 1, ..., m$) in the following form:

$$L_{ij,l} = a_{ij,l} \cdot p_{ij} + b_{ij,l} \cdot q_{ij} + c_{ij,l} \leq 0, \forall l = 1, ..., m \tag{4-11}$$

With: $L_{ij,l}$ Straight-line equation describing segment l

$a_{ij,l}, b_{ij,l}, c_{ij,l}$ Coefficients of straight-line equations

Parameters $a_{ij,l}$, $b_{ij,l}$, and $c_{ij,l}$ of each segment are obtained based on the angle $\gamma = \frac{360°}{n}$ (identical for each segment) and the maximal branch flow $s_{ij,max}$. Figure 4-2 shows an example for the approximation using a regular octagon. The apparent power limit is shown as the red dotted circle, while the boundaries of the octagon represent the eight resulting linear segments ($L_{ij,1}$ to $L_{ij,8}$).

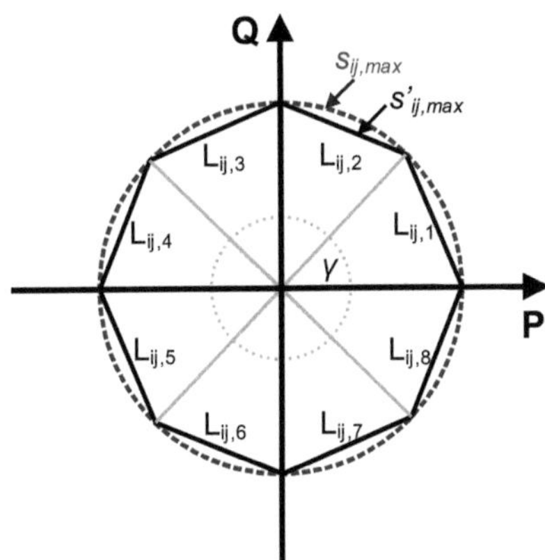

Figure 4-2: Example of piecewise linearization of branch flow constraints with n = 8 [117].

Adding (4-8) into (4-11) results in a set of linear constraints that limit the power flow over a branch. These equations depend directly on state variables u and ϑ. For the branch connecting buses i and j results in the following constraint:

$$\left(a_{ij,k} \cdot \left.\frac{\partial P_{ij}}{\partial \vartheta_i}\right|_{x_0} + b_{ij,k} \cdot \left.\frac{\partial Q_{ij}}{\partial \vartheta_i}\right|_{x_0} \right) \Delta\vartheta_i + \left(a_{ij,k} \cdot \left.\frac{\partial P_{ij}}{\partial \vartheta_j}\right|_{x_0} + b_{ij,k} \cdot \left.\frac{\partial Q_{ij}}{\partial \vartheta_j}\right|_{x_0} \right) \Delta\vartheta_j$$

$$+ \left(a_{ij,k} \cdot \left.\frac{\partial P_{ij}}{\partial u_i}\right|_{x_0} + b_{ij,k} \cdot \left.\frac{\partial Q_{ij}}{\partial u_i}\right|_{x_0} \right) \Delta u_i + \left(a_{ij,k} \cdot \left.\frac{\partial P_{ij}}{\partial u_j}\right|_{x_0} + b_{ij,k} \cdot \left.\frac{\partial Q_{ij}}{\partial u_j}\right|_{x_0} \right) \Delta u_j \qquad (4\text{-}12)$$

$$+ a_{ij,k} \cdot P_{ij}\big|_{x_0} + b_{ij,k} \cdot Q_{ij}\big|_{x_0} + c_{ij,k} \le 0$$

With:	$\Delta u_i, \Delta\vartheta_i$	Deviation of voltage angle and magnitude at bus i

The power flowing through the branch connecting the slack bus to the grid is constrained as well, i.e. by the transformer capacity or by the power line thermal limit, and this constraint needs to be included in the OPF definition. As the slack bus voltage is already included within the state vector x, it becomes trivial to create the additional constraints for the power flow through the slack bus, in the same ways as described in (4-12).

An aspect that cannot be neglected is that each segment of the piecewise linearization of the branch flow limits adds an additional constraint to the OPF model. Increasing the number of segments would help to improve the numerical quality of the resulting FOR, however, a balance between accuracy and the size of the problem needs to be found. Increasing the number of sides of the polygons has an exponential impact on the size of the OPF as it affects every branch in the grid.

4.2.6 FPU/FPG Constraints

An additional constraint is added to the OPF to describe the capability charts of the FPU/FPG. Based on the descriptions provided in Chapter 2.4, it was observed that the operational boundaries of the FPU can be represented in general through convex polygons in the complex power domain. These restrictions depict either the technical limitations of each FPU, the grid code interconnection requirements or the flexibility that the FPU can provide. This depends on the inputs given to the algorithm (see the FOR/FXOR definitions in Chapter 3.1).

Assuming a convex polygon as the capability chart of an FPU, each edge of the polygon can be described through a straight-line equation, according to (4-13). State variables $p_{FPU,f}$ and $q_{FPU,f}$ of a given FPU f are then constrained by these polygons. Following this definition, a convex linear inequalities system is obtained for the FPU, in the following format:

$$a_{FPU,f,t} \cdot p_{FPU,f} + b_{FPU,f,t} \cdot q_{FPU,f} + c_{FPU,f,t} \leq 0$$
$$\forall t = 1,..,T_f; \ \forall f \in F_i; \ \forall i \in \{B - slack\}$$

(4-13)

With: T_f Number of sides of polygon f

        $a_{FPU,f,t}, b_{FPU,f,t}, c_{FPU,f,t}$ Coefficients of straight-line equations

The coefficients $a_{FPU,f,t}$, $b_{FPU,f,t}$ and $c_{flex,f,t}$ define a straight-line equation of the tth segment of the convex capability chart of an FPU $f, \forall f \in F_i, \forall i \in B$. Each polygon f has T_f sides; therefore, each FPU adds T_f linear inequalities to the OPF. The construction of these polygons is properly described in Chapter 4.3.

4.2.7 Objective Function

The goal of the LFA method is to find the convex hull surrounding the set of valid IPF of a distribution grid. The IPF is the complex power flowing from the slack bus to the grid, defined as $p_{IPF}(x) + j \cdot q_{IPF}(x)$ and is contained in equation (4-8) [7] [22]. According to this definition, the generic objective function (4-2) is rewritten for the proposed OPF as follows [4] [22]:

$$f(x) = \gamma \cdot \left(p_{IPF}(x) + \delta \cdot q_{IPF}(x)\right)$$

(4-14)

$$p_{IPF}(x) = p_{IPF}(x_0) + \left.\frac{\partial p_{IPF}}{\partial \vartheta}\right|_{x_0} \cdot \left(\vartheta_x - \vartheta_{x_0}\right) + \left.\frac{\partial p_{IPF}}{\partial u}\right|_{x_0} \cdot \left(u_x - u_{x_0}\right)$$

(4-15)

$$q_{IPF}(x) = p_{IPF}(x_0) + \left.\frac{\partial q_{IPF}}{\partial \vartheta}\right|_{x_0} \cdot \left(\vartheta_x - \vartheta_{x_0}\right) + \left.\frac{\partial q_{IPF}}{\partial u}\right|_{x_0} \cdot \left(u_x - u_{x_0}\right)$$

(4-16)

With: p_{IPF}, q_{IPF} Interconnection Power Flow

        γ, δ Parameters defining the search direction of the objective function $f(x)$

The objective function (4-14) defines a straight line in the Cartesian plane defined by the IPF, as shown in Figure 4-3. The parameter δ defines the slope, while setting γ to -1 or 1 allows converting (4-14) into a maximization or minimization problem (with $\max f(x) = \min -f(x)$, operation valid for linear programming problems). By selecting a proper combination of γ and δ, the search direction of (4-14) can be controlled.

The optimum of the OPF, obtained after solving (4-14) with a specific combination of γ and δ, results in a specific IPF. This corresponds to one boundary point of the FOR. For each run of the optimization, a new boundary point of the FOR is obtained. The following chapter describes the developed procedure to assess the entire boundary of the FOR with reduced amount of executions of the OPF.

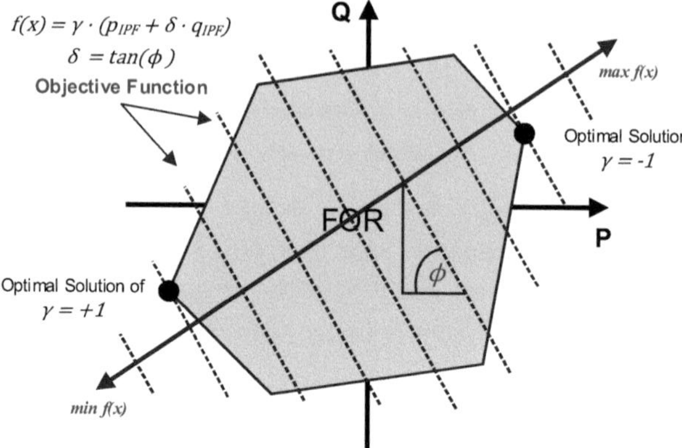

Figure 4-3: Characterization of the objective function and the FOR in the LFA method.

The assessment of grids with more than one slack bus is current subject of research, as it requires to split the IPF through the multiple slack buses, which cannot be done in a simple fashion like in this case.

4.2.8 Selection of Initial Operation Point

The initial grid operation point is obtained with a standard NR-PF with the FPU/FPG operating at nominal power. Forecasts of generation and load can be used to determine the operation points of flexible and non-flexible assets. Based on equation (4-5), any change to the operation point of the FPU (from its initial value) has impact on u and ϑ. This will cause the state vector x to drift around the initial state vector x_0. Therefore, the error of linearizing the grid model around x_0 is expected to be smaller, compared to using a flat-start as linearization point.

4.2.9 Iterative Reconstruction of FOR boundaries

The LFA algorithm offers some improvements to the methods proposed in [22] and [23]. The novelty is the subdivision of the search process in four quadrants and the use of a maximal iterations value as the break criterion. In [22] and [23], the iterative process adapts to the shape of the FOR until the difference between two boundary points is small enough. In contrast, the LFA method sets a limit for the number of OPF that are necessary to assess the FOR, emphasizing a reduction of the computational effort, without compromising the accuracy of the results. During each iteration, (4-14) is optimized using different parameters for γ and δ (cf. Figure 4-3). The proposed search procedure is illustrated in Figure 4-4.

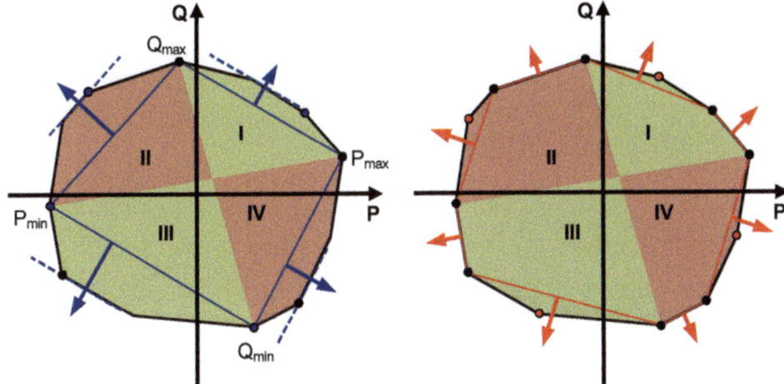

Figure 4-4: Proposed Search procedure. (left) Assessment of four extreme points, definition of search quadrants I-IV and first iteration. (right) Update of β values and identification of new boundary points for each quadrant until convergence.

The procedure follows these steps:

Step 1: Extreme boundary points P_{max}, P_{min}, Q_{max} and Q_{min} are obtained by optimizing (4-14) with $\gamma = \{-1,1\}$, and $\beta = \{0, \infty\}$.

Step 2: The intersection of the straight lines connecting the boundary points $P_{min} - P_{max}$ and $Q_{min} - Q_{max}$ define four search quadrants (I to IV in Figure 4-4). The search of the FOR boundary for each quadrant can be performed independently and in parallel.

Step 3: For each pair of adjacent points, the slope β of the straight-line connecting them is computed. Within each quadrant, β has the same sign, while γ remains constant (according to the definitions in Table 4-1). By solving (4-14) with the given values of γ and β, obtaining a new FOR vertex.

Step 4: The new boundary point is compared to previously found points for proximity. If it is located too close from an already assessed point ($|PQ_{new} - PQ_{existing}| < \varepsilon$, with ε any small value), it becomes discarded, and further iterations in this sector are interrupted.

Step 5: Steps 3 and 4 are repeated iteratively considering the newly found boundary points. A maximal amount of iterations k_{max} is defined, limiting the maximal amount of points that can be obtained within any of the four quadrants.

Step 6: Steps 3 to 5 are repeated independently for the remaining three quadrants. Table 4-1 shows the selection of the sign of β and the value of γ that need to be set in (4-14) to perform the search on each quadrant.

Step 7: The connection of the resulting points represents a piecewise linearization of the FOR of the analyzed grid.

Table 4-1: Parametrization of the objective function for each search quadrant.

Quadrant	I	II	III	IV
$sign(\delta)$	-	+	-	+
γ	-1	1	1	-1

In [23], the algorithm converges when the difference of the slope between all adjacent segments becomes smaller than a defined threshold. This can also be seen as if the difference between the slope δ of two adjacent segments were small enough. A smooth contour of the FOR can be obtained this way, however, the algorithm may have convergence issues, especially if the FOR contains sharpedges. In the LFA algorithm, a maximum amount of iterations is defined, in order to maintain an almost constant processing time. This has also the objective to avoid redundant iterations that could happen if the angle difference threshold cannot be reached. After k_{max} iterations, at most $4 + 2^{k_{max}}$ contour points can be obtained (considering the four initial extreme points). That value is not always reached, as some points are removed due to the proximity constraint and other points are not assessed as the search is interrupted.

The computation of the four initial IPF is not counted as an iteration. The resulting contour is the convex hull defined by the boundary points after the last iteration is finished. It has been observed that $k_{max} = 3$ iterations tend to be enough to obtain a representative approximation of the aggregated flexibility range (maximal 32 boundary points). Executing more iterations leads to an exponential increase in the processing time, while it would not necessarily improve the quality of the FOR.

4.2.10 On the Convexity of the FOR

Solving a power flow problem involves finding a solution of a non-linear set of equations. This can become problematic when solving an OPF, as the solution space can be non-convex. In [128], the authors demonstrated the existence of a linear approximation of the power flow problem, in which there is a linear connection between the electrical variables (u, ϑ, p, q). This approximation is capable of providing a solution to the power flow problem and can be applied to grid models with PQ buses and a single slack bus (similar to the cases analyzed in this thesis).

In the method proposed in this thesis, every component of the OPF is required to compute FOR is approximated by a set of linear equations and the constraints of the problem are certified to be convex. Based on similar premises, the authors of [127] demonstrated how the solution of a linear OPF with linear convex constraints, always results in a convex solution space. The demonstration is based on the fact that the intersection of convex areas (in this case the different constraints) results in a convex area (the FOR). The reader is encouraged to take a look into the demonstrations provided in [127] regarding this topic. The demonstration is valid to FOR obtained of linear-based OPF computations. This does not apply for exact OPF formulations with non-linear equations.

A numerical comparison between the proposed linear FOR computation approach and a RS approach is provided in Chapter 5.6, which shows the validity of the FOR concept in regular grids. However, the inherent non-linearity of electrical power systems can turn the real FOR into a non-convex solution space. This happens in similar scenarios where the NR-PF would have problems converging, e.g. grids with long branches or extreme X/R ratios. A more detailed analysis of this effect is provided in Chapter 5.4.

4.2.11 Method to Correct Linearization Error in FOR

The linearization of the power flow equations and the grid constraints in the OPF accelerates the computation of the FOR, yet this comes at a cost, as the power flow results are not as accurate compared to the solution of the non-linear model. A comparison between the linear method proposed in this work and ICPF ([22]) was done in [4]. A strong reduction in the computation time was observed, as well as some deviations in the limits of the FOR, mostly due to the error introduced with the linearization of the power flow equations.

$$\min_{x} f(x)|_{\gamma',\delta'} \rightarrow x' = \begin{bmatrix} \vartheta_i' \\ u_i' \\ p_{FPU,f}' \\ q_{FPU,f}' \end{bmatrix}, \forall f \in F_i, \forall i \in B \qquad (4\text{-}17)$$

$$NRPF(x') \Rightarrow x'' = \begin{bmatrix} \vartheta_i'' \\ u_i'' \\ p_{FPU,f}' \\ q_{FPU,f}' \end{bmatrix}, \forall f \in F_i, \forall i \in B \Rightarrow \underline{s}_{JPF}(x'') \qquad (4\text{-}18)$$

With: $NRPF$ Newton-Raphson Power Flow calculation with fix values of PQ_{FPU}

γ', δ' Parameters of (4-14) which lead to the detection of an exact IPF

x' Resulting state vector after successful optimization of $f(x)$ with parameters $\gamma = \gamma'$ and $\delta = \delta'$.

x'' Resulting state vector after NR-PF computation with operation points of the FPU obtained from x' for each FOR boundary point

This chapter describes a procedure to correct the linearization error. Each solution of (4-14) is given by a specific combination of operation points of the FPU. This is described as a state vector in (4-17). For each vertex of the FOR, a NR-PF computation is performed using the corresponding FPU operation points, from which the corrected IPF is extracted. The complex voltages in the state vector are updated, resulting in new branch flows, as described in (4-18). This is performed for all the IPF describing the boundary of the FOR. A graphical representation of the process is depicted in Figure 4-5. For each IPF computed, the operation point is corrected with a traditional NR-PF calculation. The result is the corrected FOR. A numerical analysis of this method is provided in Chapter 5.3.

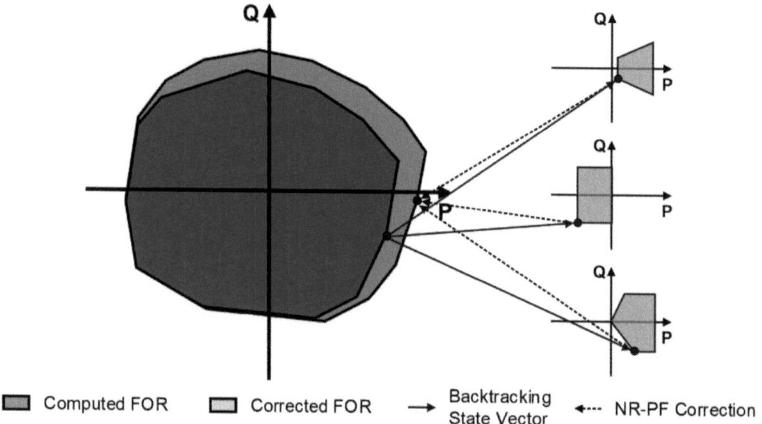

Figure 4-5: Schematic representation of proposed method to correct the linearization error in the FOR computation by performing an additional NR-PF.

4.3 Modelling of Linear FPU/FPG Boundaries

Equation (4-13) shows a generic way to include the FPU/FPG constraints into the OPF. Parameters $a_{FPU,f,t}$, $b_{FPU,f,t}$ and $c_{flex,f,t}$ are selected based on the specific characteristics of each type of FPU and the specific objectives behind the computation of the FOR. In [14], the FOR and FXOR concepts were described, meaning that the operation range of an FPU can differ, depending on different aspects of their operation. For example, in [14] the rate at which the operation point can be changed is considered to limit the FPU operation range. The definition of the boundaries is use case specific, as they could very well indicate for example the clearing of a flexibility market [40] or imposed quotas by grid operators [137].

Based on these concepts, together with the extensive literature review presented in Chapter 2.4, and previous work in [4] and [41], a set of six linearly approximated models representing different capability charts of different types of FPU is proposed in this work. An illustration of the models is given in Figure 4-6.

These models allow describing the technical limits of the corresponding FPU, based on the specific technical parameters of each device. In order to maintain the linearity of the problem, all quadratic constraints are piecewise linearized, in order to ensure numerical convergence. It is possible to integrate additional FPU models into the proposed OPF, either based on measurements or empirical values, as well as new technologies. The only restriction is that the models need to be convex. The presented models are labelled from Type 1 to Type 6, and represent typical DER found in distribution grids. An additional model is considered to represent all generic convex polygonal shapes (defined as Type 0). This helps to model FPU that do not fall into any of the aforesaid categories, e.g. FPG capability charts. A wide-ranging description of the FPU types is given in Table 4-2.

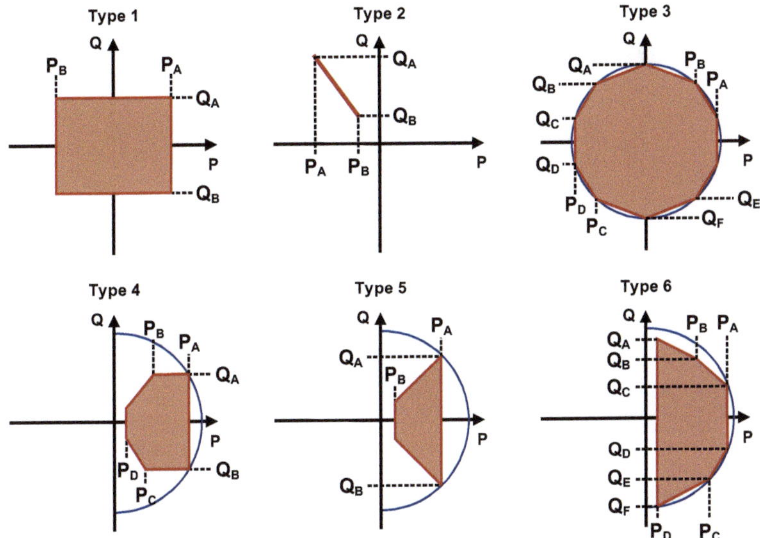

Figure 4-6: Linear constraints of six types of FPU, blue lines represent the apparent power limit.

Type 1 models imply applying (3-26) and (3-27) into the OPF, however, as the goal of approach proposed in this thesis is to consider realistic FPU operation boundaries. Based on the provided models, linear constraints are obtained from the convex capability charts of FPU, which can be derived from the parameters defined in Figure 4-6 or as a Type 0 flexibility. Parameters $P_A - P_D$ and $Q_A - Q_F$ are common to all models and are set to zero when not required.

The selection of the FPU type has a noticeable impact on the OPF, as each FPU type is defined by a different amount of inequalities and equalities, depending on the definition of the polygons. To reduce the computational complexity, inequalities of the type $p_{flex,F_i,i} \lessgtr 0$ or $q_{flex,F_i,i} \lessgtr 0$ are defined as upper/lower boundaries, while FPU of Type 2 are defined as equalities (i.e. straight-line equations). Table 4-3 provides a summary of the constraints of the OPF when adding FPU.

Table 4-2: Types of DER that can be described using the defined FPU types.

FPU Type	DER Types
Type 1	• Loads • Storage System • Reactive Power Compensation • Electrical Vehicle • Simplified Synchronous Machine
Type 2	• Load with constant $\cos(\varphi)$ • Electrical Vehicle with constant $\cos(\varphi)$ • PV with constant $\cos(\varphi)$
Type 3	• Storage system with four-quadrant inverter • Electrical Vehicle with four-quadrant inverter
Type 4	• Wind generator with reactive power limit • PV with reactive power limit • Electrical Vehicle with Q-controller
Type 5	• Wind generator with $\cos(\varphi)$ limited inverter • PV with $\cos(\varphi)$ limited inverter • Electrical Vehicle with Q-controller
Type 6	• Synchronous generator • Wind generator with D-shaped inverter • PV with D-shaped inverter
Type 0	• Generic FPU / FPG

Table 4-3: Number of constraints produced by each FPU type model divided by type.

FPU Type	Equalities	Inequalities	Upper/Lower-Boundary
Type 1	0	0	4
Type 2	1	0	2
Type 3	0	8	4
Type 4	0	2	4
Type 5	0	2	4
Type 6	0	4	2
Type 0[12]	0	n	0

---

[12] Type 0 FPU (which also describe the FPG) are modelled through n-sides convex polygons, adding n inequalities to the optimization problem.

4.4   Impact of Topology Changes on FOR Computation

The FOR is not only defined by changes in the operation points of generators, loads and storage systems, but also by the topology of the grid itself, which can change over time due to many different reasons. Distribution grids can be operated with different configurations, however, there is usually the possibility to divert the power flows by switching power lines [138] or splitting substations [139]. On the other hand, transformers can be provided with on-load tap changers, which allows for changing the entire voltage profile of a grid in discrete steps. These are some aspects that can have large repercussion in the computation of the FOR.

4.4.1   Switching Power Lines

A grid operated in open-ring topology (typical configuration for MV grids) contains switches that allow for either closing the ring, or modifying the length of the branches, by keeping the radial topology. These changes in the topology can have an impact on the voltages at the end buses of two branches in a radial network, as the impedance and loading characteristics of each branch changes. In the example of Figure 4-7, closing the ring (e.g. closing the switch S3 in Figure 4-7) would short-circuit both ends, causing both buses to be at the same potential. This procedure causes the overall voltage profile of the grid to change, consequently the transport capacity of the grid changes. The even voltages can help decongest the branch with the highest load, which can have a noticeable impact on the FOR (which depends on the line impedances). A numerical analysis of this effect is shown in later in Chapter 5.5.

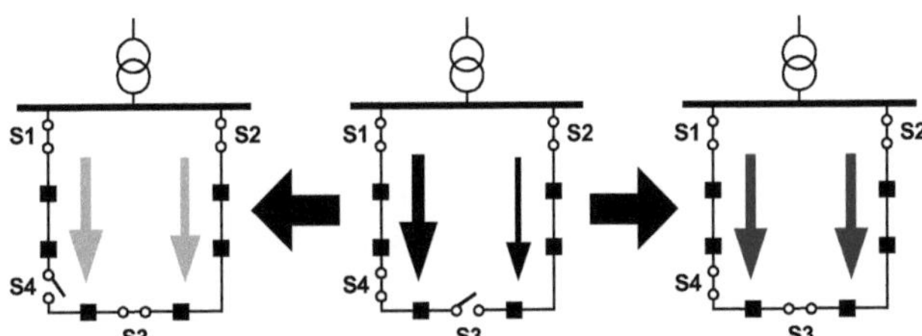

Figure 4-7: Impact of changing the grid topology on the power flows. (left) Open ring with different branch length, (center) Open ring topology, (right) Closed ring. [140]

On the other hand, substations can have a large number of possible topology configurations, which can be achieved just by reconfiguring the breakers within the premises. For example, different busbars can be connected to one another, supply power lines can be connected to the outgoing feeders individually or in parallel. A single-line diagram of a substation with multiple switching possibilities is shown in Figure 4-8.

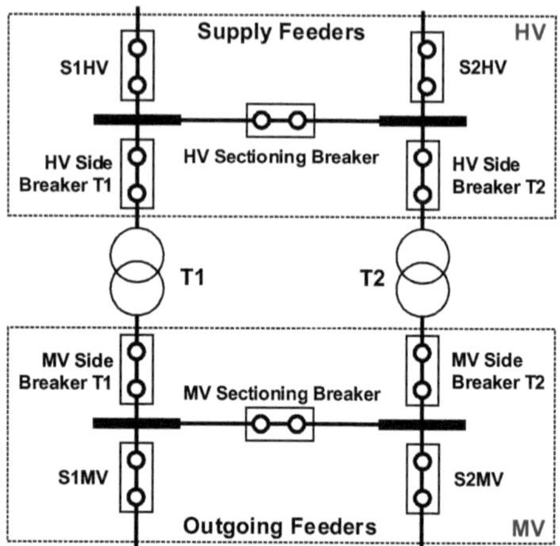

Figure 4-8: Example of possible topology configurations within a single distribution substation with two transformers and two supply lines. [139]

All these configurations have an impact on the admittance matrix Y. The connection and disconnection of power lines or transformers are reflected in Y according to (3-15) and (3-16). The switching state of the branch connecting buses i and j is defined by w_{ij} (0 if disconnected and 1 if connected).

$$\underline{y}_{ij} = \underline{y}_{ji} = y_{ij} \cdot e^{j\theta_{ij}} \cdot w_{ij}, w_{ij} \in \{0,1\} \qquad (4\text{-}19)$$

With: w_{ij} Switching state of the branch connecting buses i and j.

Changing the admittance matrix Y has a direct impact in (3-15) and (3-16), meaning that the OPF needs to be updated with help of (4-19). The reconfiguration of the busbars in a substation poses a complex challenge to the modelling of the grid, as splitting a substation into many busbars with different potential requires to add additional buses to the grid model, changing the size of Y. This results in the definition of an entire new OPF and needs to be accounted for.

A grid with k controllable switches allows for 2^k switching states; however, not all of them are practicable, and some should be avoided at all costs. To avoid leaving customers unserved, switching combinations that could cause some sections of the system to become isolated cannot be considered (e.g. simultaneously opening S1 and S2 in Figure 4-7). Other examples are switching combinations that would cause any power line to be overloaded or result in a major voltage deviation; which need to be filtered out as well. Even though this depends on the current load/generation characteristic of the grid. By filtering non-valid scenarios, the number of valid switching permutations is reduced. This is of special interest in grids with many switching possibilities, as switching combinations increase exponentially, meaning that many FOR computations would be necessary to assess all possible states of the grid.

4.4.2 OLTC Transformers

Controllable OLTC transformers can mechanically change their coil turn ratios, thus changing the voltage transformation ratio. The changes are typically discrete and are described according to a percentage of the voltage (based on the nominal turn ratio). A transformer with n possible tap positions can produce n different power flow scenarios, as the voltage profile of the downstream grid is entirely transformed with each tap position. The admittance of the transformer depends on the tap position, meaning that the admittance matrix Y is impacted by the transformation ratio t, according to (4-20) for a transformer connecting buses i and j.

$$\begin{bmatrix} y_{ii} & y_{ij} \\ y_{ji} & y_{jj} \end{bmatrix} = \begin{bmatrix} -\underline{y}_T & \underline{t} \cdot \underline{y}_T \\ \underline{t}^* \cdot \underline{y}_T & -\underline{t}^2 \cdot \underline{y}_T \end{bmatrix} \qquad (4\text{-}20)$$

With: $\underline{t}$ Complex transformation ratio of a transformer

        $\underline{y}_T$ Complex admittance of a transformer

An unfavorable selection of tap positions may cause over-/undervoltage in the grid. A proper allocation of flexibility within the grid may allow restoring the voltage levels back to acceptable limits, although this may not always be the case. Therefore, a new FOR computation needs to be performed for each tap position, cf. [4], [23] and [140].The FOR computation would allow to assess the cases where voltage issues can be solved by means of local flexibility usage. If the provided flexibility for the given period is not enough to restore the voltage, the scenario becomes unfeasible, as the aggregation algorithm will not converge.

4.5 Coupling the Linear Aggregation Method with Time-Series

Traditional flexibility aggregation methods tend to require large computation times to calculate a FOR, even for small sized grids. This limits their usability in combination with time-series as input for the calculation, as the computational time would see a dramatic increase. In this chapter, the adaptation of the proposed method to allow the assessment of the FOR considering time-series is presented (i.e. a day divided in 96 periods of 15 minutes each). First, the necessary adjustments to the FPU modeling approach is presented. Then, as proposed in [141], a procedure for sequential FOR computations is presented, together with different quantification methods. Similar approaches can be found in [7], [113], and [142], however, the novelty of this approach is the reparameterization of the FPU models coupled with time-series, as presented next.

4.5.1 Modelling of FPU Considering Time-Series

The models of the FPU capability charts introduced in Chapter 4.3 represent possible steady-state complex power values that the FPU can achieve for the period $t = t^*$. Some FPU have the capability to have an independent control of their operation point among the technically feasible operational limits. Others are dependent on variable external factors, especially RES-based generation (i.e. solar irradiation and wind speed). A PV panel cannot inject power during the night or a wind generator can only inject power within specific wind speed ranges. These effects can be incorporated in the modelling of FPU by coupling the parameters defining the polygons to time-series of the primary energy sources.

4.5.1.1 PV Generation

The injected power by a PV panel is defined by the local solar irradiation, making the active power generation $P_{PV}(t)$ time-variant. A PV generator can be modelled using an FPU of Type 2, 4 or 5, depending on the power inverter characteristics. The maximal power $P_{PV,max}$ is defined by the installed capacity of the PV generator, which is a static characteristic of each PV. For all three FPU types, the parameter $P_{PV}(t)$ defines the active power that the PV generator can inject due to the solar irradiation in a given period, leading to the redefinition of the time-variant parameter P_A as in (4-21). A schematic representation of the time-variant capability chart of a PV defined as a Type 5 FPU is shown in Figure 4-9.

$$P_A(t) = P_{PV}(t) \leq P_{PV,max} \tag{4-21}$$

With: $P_{PV}(t)$ Injected active power by the PV generator at time t

          $P_{PV,max}$ Rated capacity of PV generator

4.5.1.2 Wind Generation

The active power injection of wind generators depends on the on-site wind speed. This dependency is represented in $P_{WG}(t)$, the active power generation of the wind turbine over time. Similar to the PV case, the time-series are coupled to the FPU model (either Type 1, 4 or 5) according to (4-22), where the maximal power is equivalent to the rated active power of the generator ($P_{max} = P_{nom}$). A schematic representation of the proposed model of the flexibility provision of a WG defined as a Type 4 FPU over an entire day is shown in Figure 4-10.

$$P_A(t) = P_{WG}(t) \le P_{WG,max} \tag{4-22}$$

With: $P_{WG}(t)$ Injected active power by WG at time t

          $P_{WG,max}$ Rated capacity of WG

          $SOC(t)$ SOC of the storage system at time t

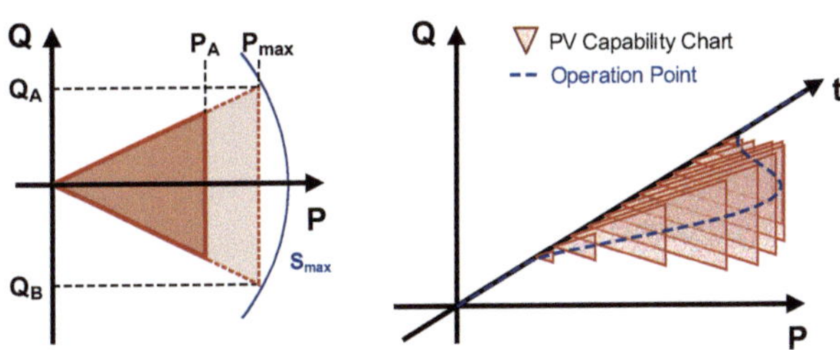

Figure 4-9: Time-variant capability chart of PV generator defined as Type 5 FPU. [141]

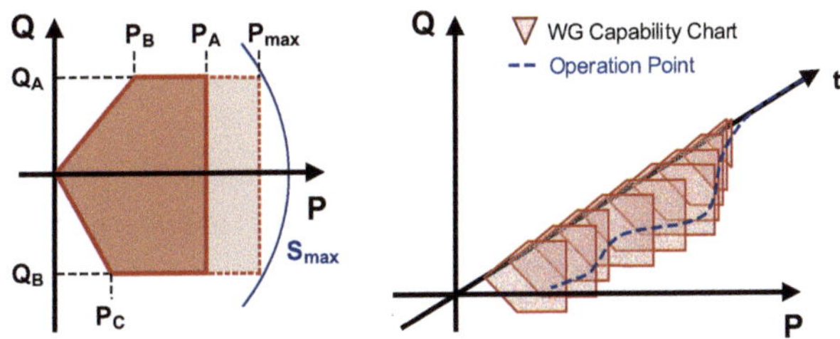

Figure 4-10: Time-variant capability chart of wind generator defined as Type 4 FPU. [141]

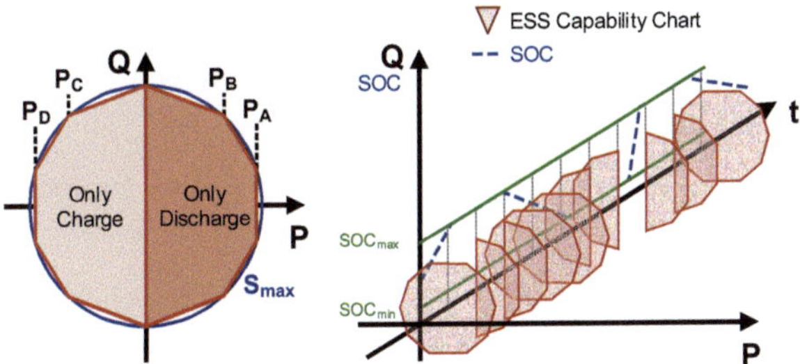

Figure 4-11: Variation of flexibility provision of ESS over time. The size of the polygon changes only according to the SOC limits. [141]

4.5.1.3   Storage Systems

The operation of storage systems depends on two main factors, the state-of-charge (SOC) and the maximal charging/discharging powers, which can be symmetrical or asymmetrical. Here, the capability chart of a storage system is assumed as time-invariant as the operation is controlled through a battery management system (BMS). Depending on the reactive power capability provided by the power inverter, Type 1, 2, and 6 models can be used. When the SOC reaches a maximum (SOC_{max}) or minimum (SOC_{min}) level, further charge or discharge is not possible, thus the flexibility provision chart is constrained to avoid these cases. The SOC needs to be considered as a state variable in the optimization problem, otherwise the intertemporal behavior of the storage system cannot be assessed.

$$if\ SOC(t) = \begin{cases} SOC_{max} \Rightarrow P_A(t) = P_B(t) = 0 \\ SOC_{min} \Rightarrow P_C(t) = P_D(t) = 0 \end{cases} \qquad (4\text{-}23)$$

With: SOC_{max}, SOC_{min} Max. and min. SOC allowed by the storage system

4.5.2 Sequential FOR Computation Using Time-Series

Based on the time-variable models of the FPU, a time-based computation of the FOR is possible. A sequential aggregation algorithm was proposed in [141], in which the FOR is computed independently for each time-step following the procedure described in Figure 4-12. It is a five-step algorithm, which include the adaptation of the FPU models, the computation of the FOR and the summation of the results. The steps shown Figure 4-12 are detailed next:

Step 1: Define grid model, assign FPU/FPG and define interconnection point to overlayered grid where the IPF is to be calculated.

Step 2: Modelling of the FPU/FPG capability charts according to the specifications shown in Figure 4-6.

Step 3: Coupling time-series to time-variable FPU based on historical data, synthetic data or forecasts, depending on the use case.

Step 4: Sequential application of the aggregation algorithm for each time-step, providing a new $FOR(t)$ each time-step t.

Step 5: The obtained FOR are analyzed, considering their evolution over time or the distribution of the overlapping of the obtained FOR in complex power domain.

The process assumes that there are no time-dependent variables, therefore, each computation is independent and the process could easily be parallelized, which is of enormous help in order to reduce the computation time, as is shown later in

Chapter 6.1. If the SOC of storage systems is to be considered, a parallel process to estimate its value is necessary. The development of such processes is not in the scope of this work as they can be very specific for each use case. Therefore, for sake of simplicity, the SOC is not considered as an independent variable in any of the forthcoming analysis.

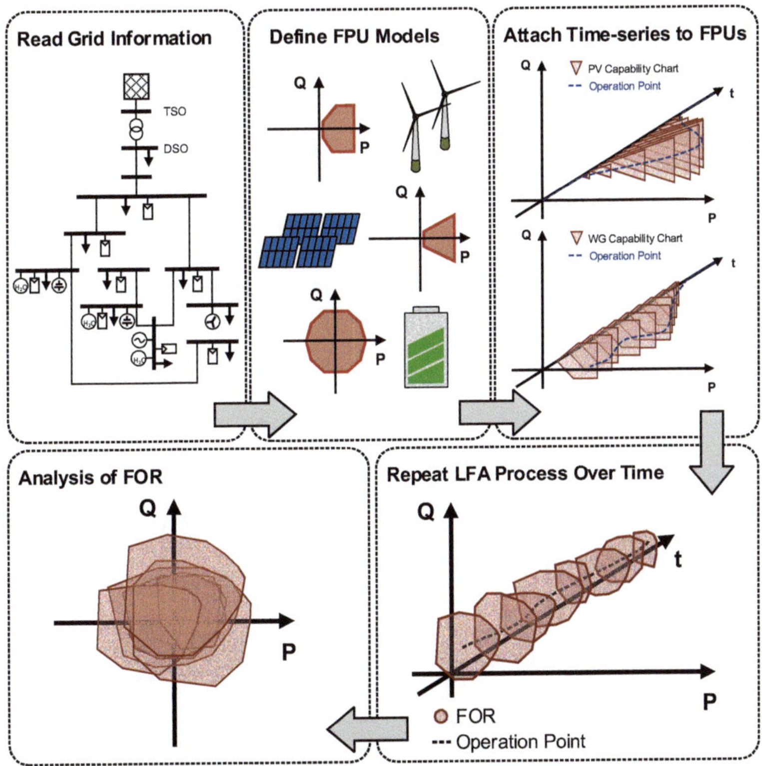

Figure 4-12: Schematic representation of the LFA aggregation process at the TSO/DSO interface combine with active power time-series. [141]

5   Validation of Linear Flexibility Aggregation Method

This chapter shows different quality assessment methods for the proposed LFA algorithm, which allows the computation of the FOR of radial distribution grids. First, an overview of the grid models used for the various analysis is provided. Then, the error added by linearized power flow equations is quantified, so its impact in the quality of the FOR can be estimated. Later, the convexity of the linear OPF is demonstrated, compared to the non-linear power flow equations. Subsequently, the impact of the grid topology in the FOR is analyzed in detail. Finally, a comparison of the LFA method against similar algorithms is provided.

5.1   Grid Models

The use of the LFA algorithm is analyzed using a collection of distribution grid models, which are introduced in this chapter. The models are based on one hand on the well-known MV/LV CIGRE European testbench [143], and on the other hand, on the abstraction of a real urban MV/HV distribution grid. The topologies and general specification of the models are presented here, while specific details regarding grid impedances, load sized and FPU parametrization (based on the models described in Figure 4-6) can be found in Appendix B.

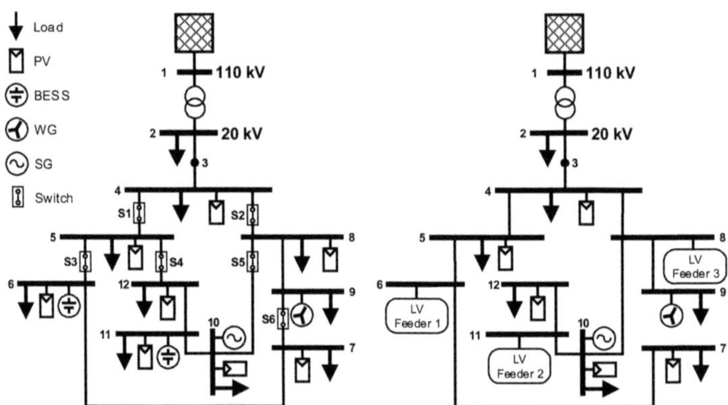

Figure 5-1: Adapted CIGRE European MV distribution grid. (left) Including circuit breakers [140], (right) Adaptation with underlayered LV feeders [41] [144].

5.1.1 CIGRE MV Grid

The CIGRE MV grid proposed in [143] was adapted to consider only the largest of the two feeders. Figure 5-1 shows two variations of the grid model. The first one includes additional switches, in order to allow assessing the impact of topology changes in the FOR. Figure 5-1 (left) shows the standard switching state of the grid (S4 and S6 are open), which is used in all simulations, unless the contrary is indicated. The second variation includes three detailed underlayered LV feeders, which are described in the next chapter. The grid models consider all loads to be inflexible, while all generators and storage systems are modelled as an FPU. The load attached to each bus represents the aggregated residual load at that feeder, while the single generators and storage systems are modelled in more detail, following the definitions given in [144].

5.1.2 CIGRE LV Feeders

The CIGRE testbench of [144] proposes three different LV feeders, representing three different types of customers, based on the European perspective. The three feeders represent a commercial (LV Feeder 1), an industrial (LV Feeder 2) and a residential (LV Feeder 3) area. The topology of the feeders is shown in Figure 5-2, while the parametrization of the FPU can be found in Appendix C.

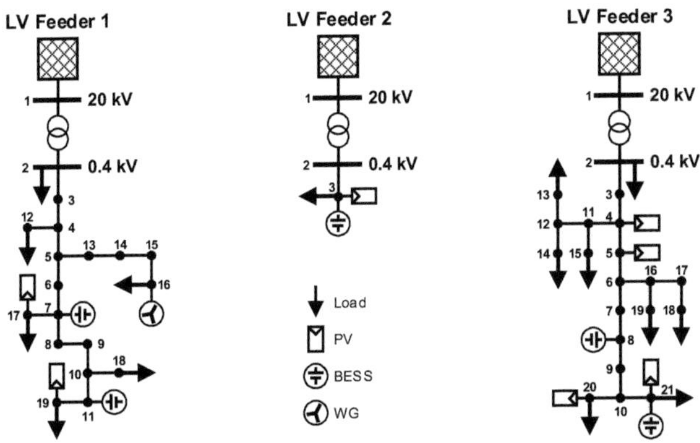

Figure 5-2: Adapted CIGRE European LV feeders. [144]

5.1.3 Real Urban Distribution Grid

A model based on a real urban distribution grid is used to analyze the benefits of the proposed aggregation method, when larger numbers of FPU and FPG are considered. This particular distribution grid covers the 110, 20 and 0.4 kV voltage levels, however, for modelling purposes, only the 110 kV grid and three of the seven 20 kV feeders are modelled in detail. The meshed topology of the 110 kV grid can be seen in Figure 5-3, where the connection point to the detailed 20 kV feeders are specified. The external grid signifies the interconnection to the TSO.

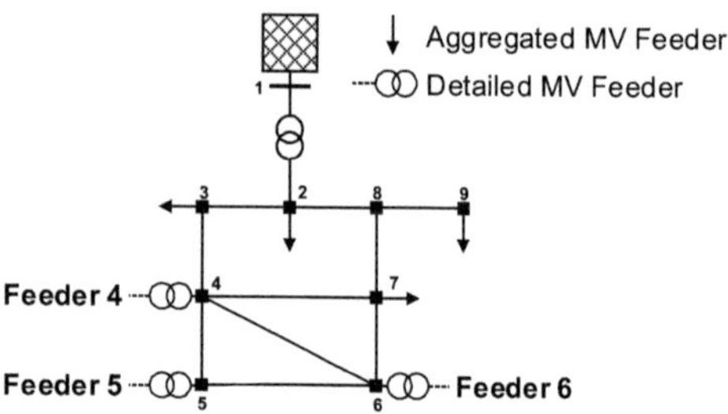

Figure 5-3: 9-Bus 110 kV meshed grid with interconnection to three detailed MV feeders.

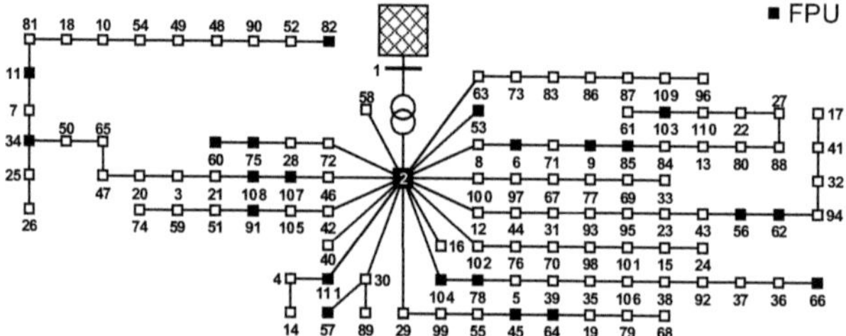

Figure 5-4: 111-Bus 20 kV Radial Feeder 4.

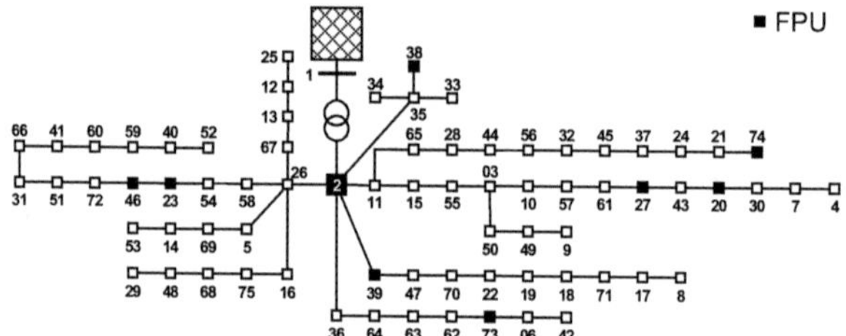

Figure 5-5: 74-Bus 20 kV Radial Feeder 5.

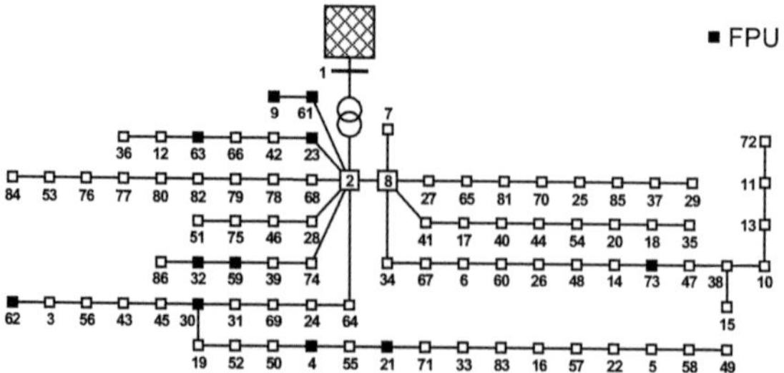

Figure 5-6: 86-Bus 20 kV Radial Feeder 6.

There are no generators connected directly to the 110 kV grid, whereas the three 20 kV feeders (shown in Figure 5-4, Figure 5-5 and Figure 5-6) have DER of different types and sizes connected to them. The specifications of the FPU connected to the 20kV feeders are described in appendix C.

5.2 Validation of Linear Power Flow Model

The main characteristic of the LFA method is the definition of a linear OPF, including a linear power flow equations system, as was detailed in previous chapters. A comparison of the LFA with the non-linear ICPF algorithms was performed in [4], where the resulting FOR showed some discrepancies associated with the error added in the linear approach. This chapter focusses on analyzing the magnitude of the error and its conceivable impact in the FOR. The analysis is based on the procedure described in [145], which analyses a similar linear power flow method applied to a LV grid, obtaining differences in the voltage magnitude smaller than 0.01 p.u., while the maximal angle difference was 0.13°. In this case, the linearization error is analyzed on the CIGRE MV grid.

For the analysis, all generators and storage systems connected to buses 3 to 12 in the grid model of Figure 5-1 are removed and replaced by single FPU of Type 1. The capability chart of the FPU is limited between {-1,1} MW/MVAr and random complex power operation points are selected within these boundaries using two independent uniform distributions (for active and reactive power).

After selecting a random operation point for each FPU, a load flow computation is performed. On one side, using the proposed linear power flow model described in (4-6) and (4-8), and on the other side, using a classical NR-PF calculation[13]. In total, 5 million power flow calculations were performed using both methods, with the same inputs each time. After each repetition, the voltage magnitudes and angles and the power flowing through the branches ($p_{ij} + j \cdot q_{ij}$) obtained from both methods were compared.

The boxplots in Figure 5-7 show the distribution of the differences in both the voltage magnitude and angle between the linear and the exact power flow approaches. The difference between both models is computed according to:

$$\Delta x = x_{NR-PF} - x_{lin} \tag{5-1}$$

With: x_{NR-PF} Variable obtained from the NR-PF computation

          x_{lin} Variable obtained from the linear load flow computation

---

[13] All NR-PF computations are performed using MATPOWER [156]

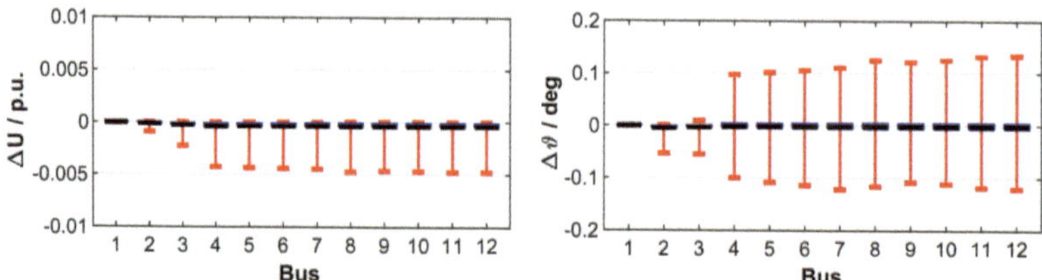

Figure 5-7: *Linearization error in the bus voltage angle and magnitude. (left) Voltage magnitude, (right) Voltage angle.*

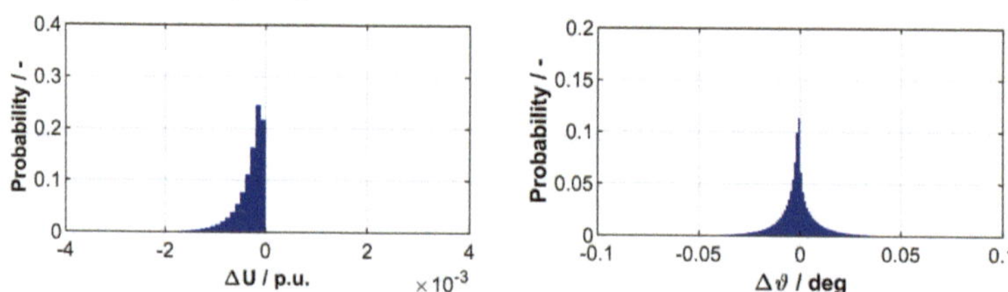

Figure 5-8: *Histogram of the error in the voltage angle and magnitude of all buses. (left) Voltage magnitude, (right) Voltage angle.*

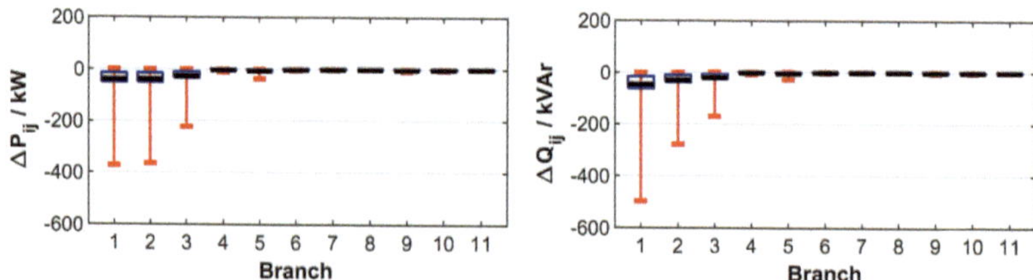

Figure 5-9: *Linearization error in the active and reactive branch flow. (left) Active power flow, (right) Reactive power flow.*

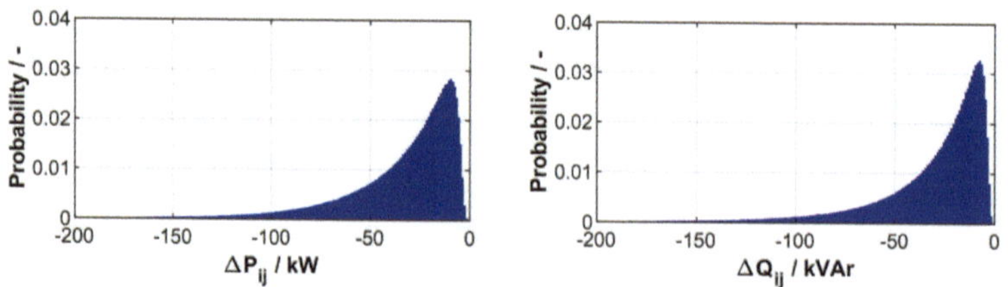

Figure 5-10: *Histogram of the error in the branch flow among branches 1 to 3. (left) Active power flow, (right) Reactive power flow.*

In the analyzed grid, the voltage difference at each bus increases with the distance to the slack bus. The results of the NR-PF are used as reference, therefore, the linear approach provides larger voltage magnitudes than the NR-PF method, while the angle difference goes in both directions. The maximal deviation of the voltage magnitude is bounded by 0.005 p.u. and the maximal angle difference is bounded by 0.133°. The probability distribution of the errors is shown in Figure 5-7 (slack bus is excluded because it remains constant).

An error in the voltage has the consequence of an error in the branch flow computation, which is subject of this chapter. It can be observed in Figure 5-9 that the largest error is obtained among branches 1 to 3; the HV/MV transformer, and the power lines connecting buses 2-3 and 3-4 respectively (see Figure 5-1). This is the interconnection point to the overlayered grid, meaning that it is expected to be the most loaded branch. The maximal difference is observed at the transformer, with 366 kW and 500 KVAr, however, the mean error is just 38 kW. Compared to the mean power flow through the transformer of 20.69 MVA, becomes almost neglectable. Figure 5-10 shows the probability distribution of the error for the three aforementioned branches; showing similar behaviors in the cases of active and reactive power. The 99th percentile of the difference in the three branches is located at 187 kW and 318 kVAr, bounding the maximal expected errors.

The outcome is that the linearization approach introduces an error into the power flow computation, whereas for the computation of the bus voltages it is not worrisome. Nonetheless, the error in the branch flows can have an impact on the computation of the FOR, which characterizes a collection of branch flows, therefore, it is of interest to quantify the magnitude of that error. Under some operation conditions, the error in the FOR could be in the order of 200 kW. As an absolute value, this cannot be overlooked, however, a different interpretation can be given when compared to the overall size of the FOR. Problems may arise when the dimensions of the FOR and the linearization error are of similar magnitude, leading to possible inaccuracies in the FOR, especially as grid constraints may be poorly assessed. Therefore, the accuracy of the FOR needs to be validated.

5.3   FOR Linearization Error Correction in LFA

The error introduced by the linear power flow model was described in the previous chapter, while Chapter 4.2.11 described a method to cope with this error in the computation of the FOR. This chapter demonstrates how the proposed correction method operates, as well as its impact on the resulting FOR. The scheme shown in Figure 4-5 is implemented as a complement to the LFA algorithm, where each vertex of the FOR computed using the LFA approach is recomputed using a NR-PF. The FOR vertices correspond to IPF originating from specific combinations of FPU operation points. If the operation points of the FPU for each of the IPF is used as input for the NR-PF computation, the IPF can be corrected. The expected result is a new IPF slightly shifted from the original one. The correction method is analyzed and compared to a brute-force computation of the FOR using NR-PF, in order to have a comparison to a "real" FOR. The LV Feeder 2 of Figure 5-2 and the MV grid of Figure 5-1 are employed in the study

In a brute-force computation of the FOR, a separate NR-PF calculation is performed for every possible combination of operation points as allowed by the FPU capability charts. Such an analysis is only possible with a limited number of FPU, otherwise the computational complexity explodes. In order to reduce the possible combinations, as they are mathematically infinite, a homogeneous lattice is defined for each FPU. This allows reducing the number of combinations, while taking the entirety of the capability charts into consideration.

The number of FPU in the analyzed grids was also decreased. For the LV feeder, only one FPU of type 3 and one of type 5 are considered, while on the MV grid, just three FPU were considered (one of each type 4, 5 and 6). A NR-PF is computed for each combination of operations points. Analogous to a RS approach, the validity of the power flows is analyzed and non-valid IPF (due to grid constraint violations) are neglected. The convex hull containing the valid IPF is defined as the FOR. This would be the most accurate method to get the "real" FOR of a power system, yet is extremely unpractical in most larger grids. Figure 5-11 shows the FOR obtained using the LFA method (with and without the suggested correction) and brute-force NR-PF.

In the selected scenarios, the FOR resulting from the brute-force approach shows almost no deviation compared to both LFA results. This is especially true, when the FOR is not limited by the grid constraints, as in the MV grid case (Figure 5-11, left). However, as can be seen in the case of the LV grid (Figure 5-11, right), the limitations imposed by the transformer (showing as the rounded edge on the right side of the FOR), would cause the correction method to fail.

The expected linearization error can have an impact in the resulting FOR, to which the correction method of Chapter 4.2.11 was developed. This statement cannot, however, be generalized for all types of grids, as the analysis only covers two rather small scenarios. Yet, the obtained results showcase the proposed correction method as a promising solution, still, a more exhaustive analysis is necessary to assess the impact of the linearization error in the computed FOR.

The analysis is extended to study the impact of the grid impedance in the correction method. The behavior of the power flow equations becomes erratic in power systems with large impedances (e.g. a radial grid with extremely long power lines). Therefore, the analysis previously described for the MV grid is repeated, only this time with an increase of the branch impedances of 200% and 400%. This follows the purpose of magnifying the linearization error. The R/X ratios of all branches remain unchanged in all cases.

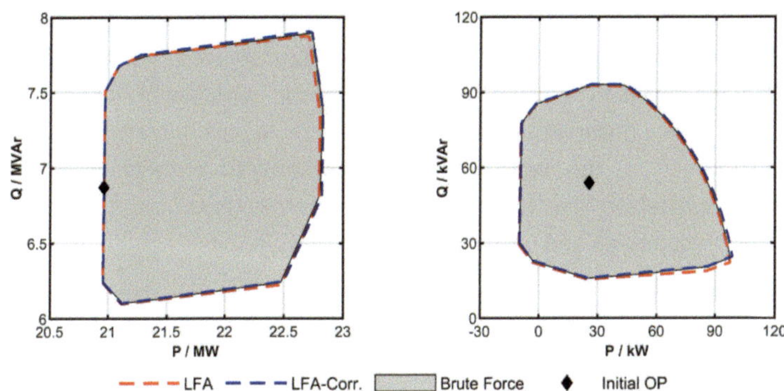

Figure 5-11: Demonstration of LFA with FOR correction method with brute-force NR-PF computations. (left) MV Grid, (right) LV Feeder 2.

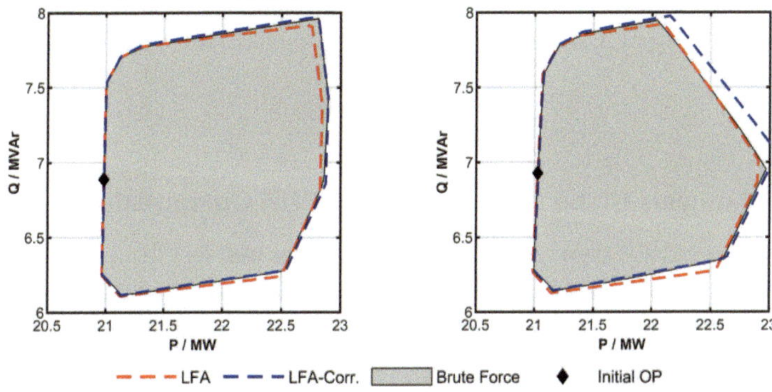

Figure 5-12: FOR correction method for the LFA compared to brute-force NR-PF computations in the MV Grid with increasing branch impedance. (left) 200% branch impedance, (right) 400% branch impedance.

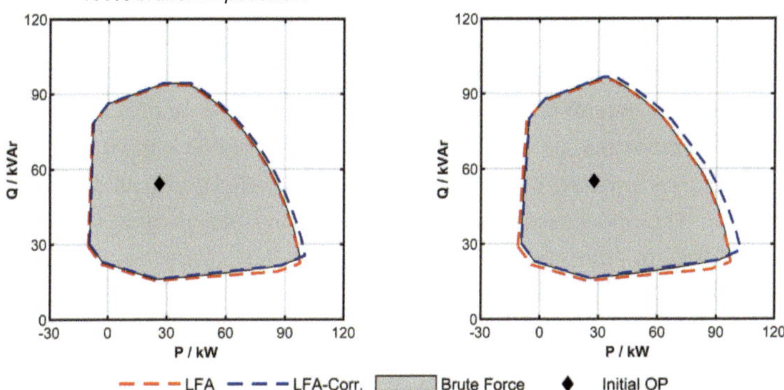

Figure 5-13: FOR correction method for the LFA compared to brute-force NR-PF computations in the LV feeder 2 with increasing branch impedance. (left) 200% branch impedance, (right) 400% branch impedance.

Figure 5-12 (MV Grid) and Figure 5-13 (LV Feeder 2) show that the correction method plays a larger role as the impedance is increased, with a noticeable growth in the deviation between the corrected and uncorrected FOR coming from the LFA. It is also observable that neither the corrected nor the uncorrected FOR of the LFA matches the "real" FOR (obtained through a brute-force approach) entirely. Nonetheless, the "real" FOR can be approximated from a combination between the corrected and uncorrected FOR.

Based on this example, defining the resulting FOR of the LFA method as the intersection of both corrected and uncorrected FOR is seen as the best option in this case. This has the benefits of correcting the linearization error and accounting for grid constraints properly. It is important to prevent declaring unfeasible IPFs as feasible, therefore, and underestimation of the FOR should be favored over an overestimation of it. This results in the following redefinition of the FOR obtained from the LFA method, which is used in the subsequent analyses as well.

$$FOR = FOR_{LFA} \cap FOR_{LFA_{Corr}} \tag{5-2}$$

5.4 Analysis of the Convexity of the FOR Computation

The impact of DER parametrization in the grid operation has been already analyzed. One studied aspect was how the grid conditions can impact the capability chart of the plant. What was observed in [146] is that "the plant equipment (cables, transformers etc.) can lead to a mismatch between the operation points of the single DG units and the operation points of the entire DG plant at the NCP (Network Connection Point)". An example of a non-convex FOR is also provided in [146], resulting from a convex FPU capability chart, which bring in mind the observations provided in [84] for a wind farm and in [147] for a PV system.

A critical issue with the FOR is that its "real" shape is initially unknown and any computation of it, regardless of the method, just provides an approximation of the shape. RS approaches, as shown in Chapter 3.2.3, provide a good overview of the "real" FOR. However, this is only feasible in smaller grids, as discussed in Chapter 5.3. The grid losses, grid constraints and the non-linearity of the power flow equations are three key factors that shape the FOR. Here, the impact of the non-linearity of the power flow equations on the shape of FOR is demonstrated in a 3-bus grid with a single FPU (Figure 5-14). Under certain conditions, the FOR can become non-convex. This chapter describes how the LFA method can cope with the non-linearity and the non-convexity of the power flow equations.

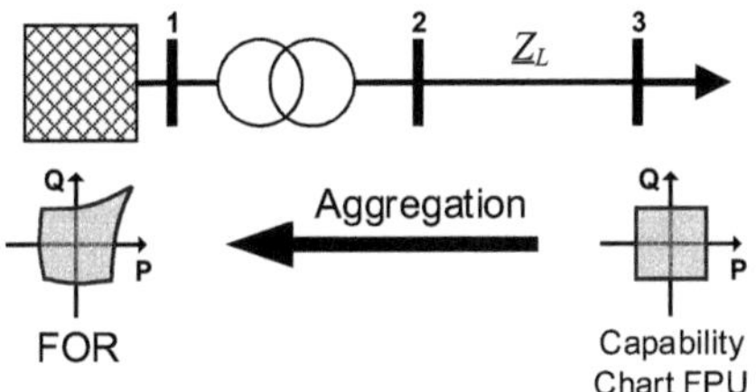

Figure 5-14: 3-Bus grid model with power line of variable length defined to evaluate the impact of the grid non-linearity in the FOR.

The FOR is computed simultaneously with the LFA and the RS methods (with 10 million samples). The power line connecting buses 2 and 3 has an impedance $\underline{Z}_L = (0{,}266 + j0{,}203)\,{}^{\Omega}/_{km}$. The length of the line is gradually increased and the resulting FOR are compared. Voltage limits are set to 0.9 and 1.1 p.u. at all time.

The branch flow constraints are considered in the analysis; however, they are unreachable in these scenarios, having no impact in the results. The FPU is defined as Type 1 with ±250 kW/kVAr limits, as shown in Figure 5-15.

The non-convexity of the FOR becomes evident as the impedance increases, especially due to the impact of the voltage constraints, evidencing the correctness of the mathematical descriptions of this peculiar effect in [10] and [109]. In the shown cases, not only the overall shape of the IPF loses its squareness, but also the grid constraints start appearing, resembling the model described in Chapter 3.2.1. Yet, a significant impact is only observed when the power line is extremely long, in this case >200 km, which is unrealistic at the MV level. A noticeable aspect is that the FOR obtained using the LFA method (black dashed line) provides a convex polygon contained within the "real" FOR (green area).

The experiment is repeated with larger FPU (±2.5 MW/MVAr). If the FPU increases its flexibility output, the threshold distance at which the FOR becomes non-convex is reduced, as in Figure 5-16. The voltage restrictions begin to be noticeable at 20 km, much shorter distance than in the previous case.

The analysis is repeated for LV grids, in the 3-bus constellation with an impedance $\underline{Z}_L = (0{,}126 + j0{,}069)\,{}^{\Omega}/_{km}$ and the FPU limited to ±150 kW/KVAr. The cable length is again gradually increased until reaching a length of 250 m. Figure 5-17 shows a similar behavior than in the MV grid, at a distance of at least 100 m.

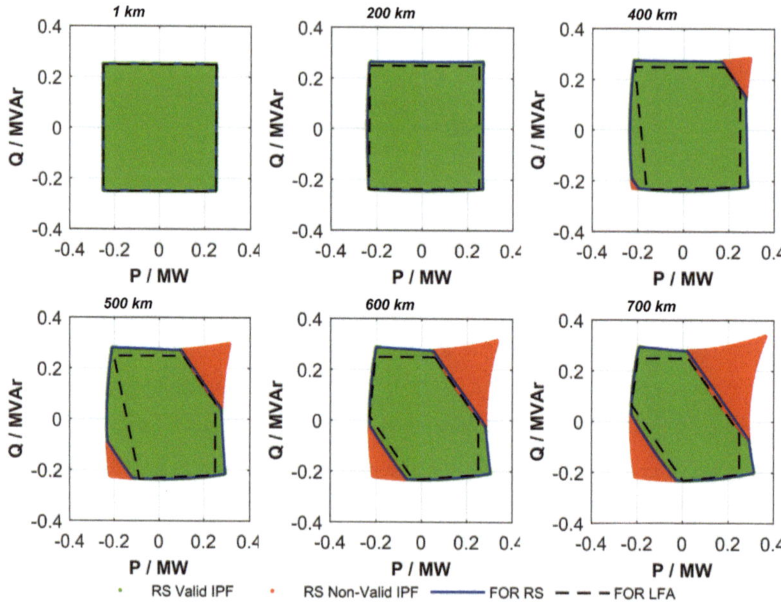

Figure 5-15: FOR with different power line lengths in a 20 kV grid with ±250 kW/kVAr flexibility.

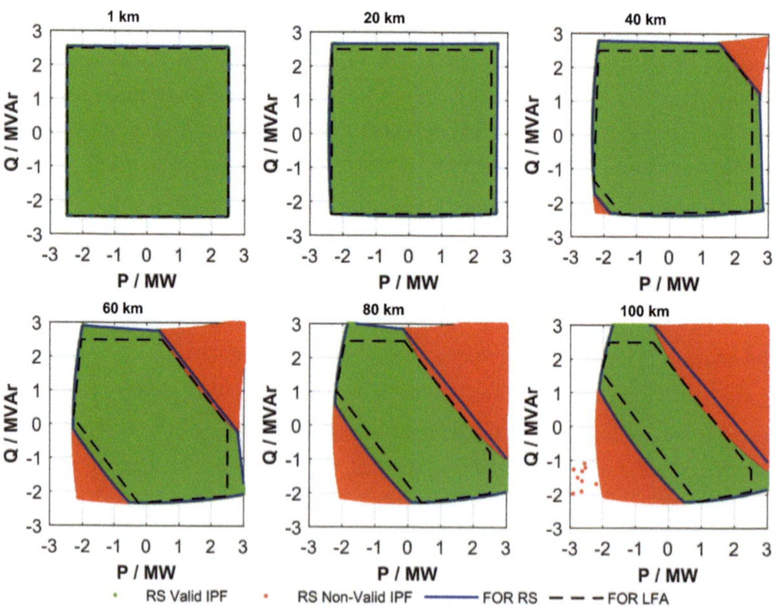

Figure 5-16: FOR with different power line lengths in a 20 kV grid with ±2.5 MW/MVAr flexibility.

Two main conclusions can be drawn from these chapters. First, that the shape of the FOR is strongly dependent on the grid topology and impedance, as well as the FPU capability chart. The transport of the full FPU capability over a single power line was analyzed here. This can change, however, if instead of the 3-bus model an n-bus model with the same total impedance and the FPU capability is shared among all the buses. The second conclusion is that the LFA method always provides a convex FOR, even when the "real" FOR is not convex. At a given impedance, the linear FOR begins to underestimate the "real" one, which is not ideal. Yet, this is preferable than the opposing scenario, where the FOR is overestimated, meaning that non-valid operation points are declared as valid. The non-convexity of the "real" FOR becomes obvious only in extreme scenarios, where grid constraints are already being stressed. This increases the error in the computation of the linear FOR. Special attention is required while applying the proposed method in grids with unusually long branches, e.g. a rural grid with large distances between consumers; as well as meshed grids.

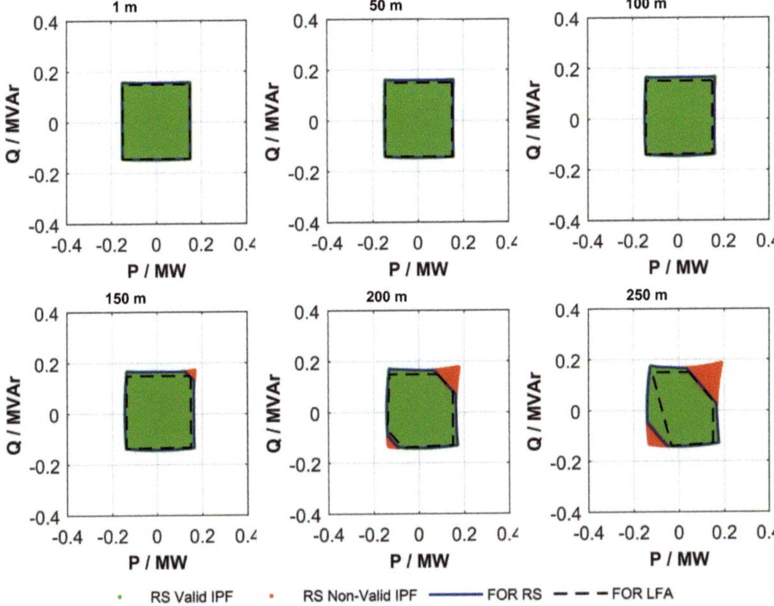

Figure 5-17: FOR with different power line lengths in a 0.4 kV grid with ±150 MW/MVAr flexibility.

5.5 Impact of Grid Topology and Constraints in the FOR

A radial distribution grid can provide different amounts of flexibility to an overlayered grid depending on the way the power lines and transformers are operated.

The grid topology is integrated in the LFA calculation in the matrix J in (4-5). New topologies can be generated by reconfiguring the switching states or by changing tap positions in an OLTC transformer, resulting in new matrices J. This means that each individual topology would require an independent computation of the FOR using the LFA method. On other hand, the selected grid constraints also have an impact on the LFA method, acting directly in the OPF definition. These two aspects are analyzed in this chapter, where the impact on the computed FOR caused by modifying the grid topology is analyzed; the methodology presented in [140] is followed. The CIGRE MV grid model shown in Chapter 5.1.1 is used as reference.

The action of switching power lines on and off and its impact in the grid power flows was already discussed in Chapter 4.4. This concept is applied in conjunction to the LFA method to verify the actual impact on the FOR by modifying the grid topology. When defining a new topology for the grid, depending on the motivation behind the requirement (e.g. fault isolation, losses reduction, equipment malfunction or maintenance), it is important to keep the number of unserved customers at zero, or at least to minimize it. For example, in Figure 5-1, a bad scenario would be to open switches S4 and S5 simultaneously, leading to a blackout of all customers connected to buses 10-12. This might be understandable in the case of fault isolation, but not in the case of grid reconfiguration due to congestion management or technical losses reduction measures. Initially, only switches S1 to S6 are considered, giving a total of 64 possible switching combinations. By applying a scenario reduction based on [140], the feasible scenarios, the ones that ensure the supply of all customers, are reduced to 19 (29,7% of the total scenarios).

For each valid scenario, a FOR is computed using the LFA method. Figure 5-18 (right side) shows the superposition of the resulting FOR obtained with each valid topology. The color map shows the probability that an IPF is feasible in all aforementioned scenarios. This means that some IPF could be achieved regardless of the selected topology (red), while others can only be reached in some very specific cases (blue), or not at all (white). In general, a fully meshed grid, i.e. all switches closed, would provide the largest FOR, while in purely radial grids the FOR begins to be affected by the grid constraints (lowest voltage limit in this example). Power line capacities do not impact the FOR, just the transformer maximal loading, as can be observed by comparing the two graphics in Figure 5-18, where a large area of the FOR is limited by the 25MVA capacity of the transformer.

Changing the grid constraints can have a large impact in the FOR. Figure 5-19 (left) shows the impact of applying stricter voltage constraints to the grid, which in this example has a strong impact, as the initial IPF (black marker) is already unfeasible and could not be operated safely under any topology. Through this analysis, it can be realized that the operation of the grid needs to be corrected, in order to bring the voltage back to permissible limits. Figure 5-19 (right) shows the impact of reducing the power lines capacity to a 35% of the nominal value.

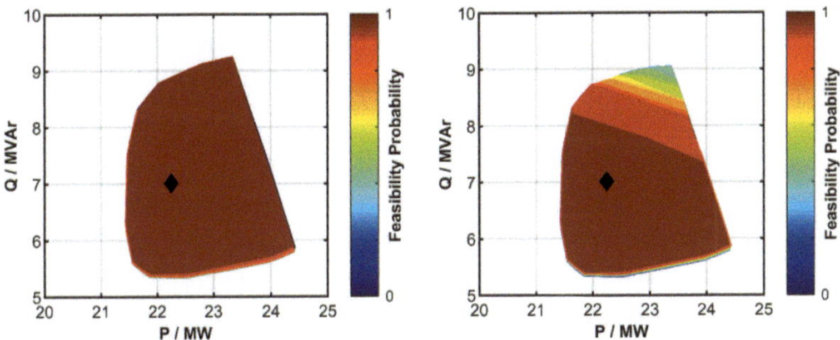

Figure 5-18: Superimposed FOR of valid switching states of S1-S6. (left) Without considering grid constraints, (right) With grid constraints ($U_{nom}\pm10\%$).

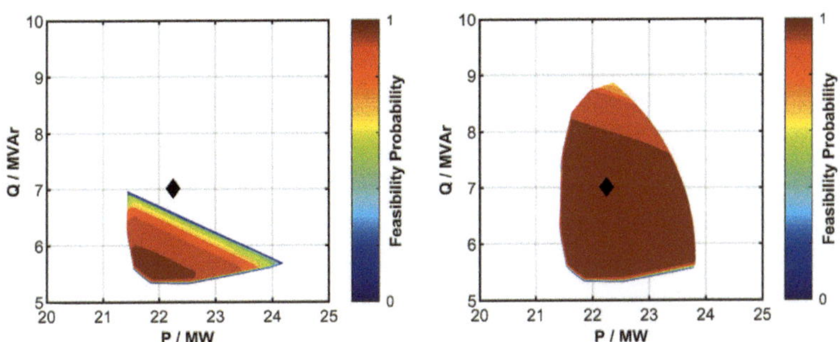

Figure 5-19: Superimposed FOR of valid switching states in S1-S6. (left) Reduced voltage constraints $U_{nom} \pm 5\%$, (right) Reduced branch flow constraints $0.35*S_{max}$.

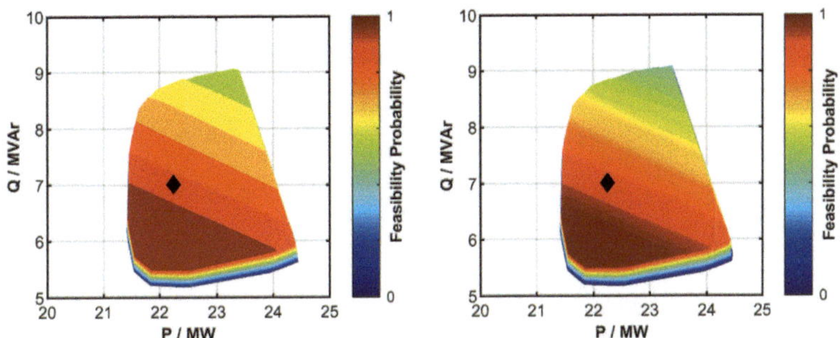

Figure 5-20: Superimposed FOR with changing tap positions of OLTC transformer. (left) Default switching scenario (S4 & S6 open), (right) All valid switching scenarios.

The transformer tap changer plays a crucial role in the downstream voltage of a distribution feeder. For this analysis, the tap positions of the MV/HV transformer connecting buses 1 and 2 are changed between -5% and +5% in 1% steps. In Figure 5-20 (left), the impact of the tap changer can be clearly observed (each color step reflects a tap position). When the grid voltage is reduced and the low-voltage limit of 0.9 p.u. is reached, the area becomes constrained in the upper right side (due to excessive reactive power absorption from the FPU). Figure 5-20 (right) shows all possible combinations between switching states and tap positions in a type of superposition of both characteristics.

The flexibility provision of a grid is strongly related to the technical and operational characteristics of that grid, as well as all pertinent safety margins. The FOR computation will provide helpful information only if the grid is operating in a valid state, or in a scenario where a valid operation state can be reached with the use of local flexibility. Under some grid configurations, the LFA algorithm will fail to converge, meaning that the ADN is not able to provide flexibility to the overlayered grid, since it cannot ensure its own stable operation first.

5.6 Comparison Between LFA and RS Approaches

The previous chapter focused on proofing the quality of the proposed LFA method to compute the FOR, especially considering the linearization error. This chapter focuses on comparing the LFA method to similar random sampling approaches, in order to evaluate how much the assessed FOR differs from the "real" FOR. Therefore, the FOR is computed simultaneously using the described LFA and RS methods, the last one using PDF described in Chapter 3.2.3. The impact on the FOR of the proposed PDF is analyzed, and the results are compared with the ones of the LFA method. The comparison is done following the methodology proposed in [108] and [119]. Initially, the CIGRE MV grid (Figure 5-1) is analyzed, while a larger grid is considered in the following chapter.

$$similarity(FOR_1, FOR_2) = \frac{area(FOR_1 \cap FOR_2)}{area(FOR_1 \cup FOR_2)} \cdot 100\% \qquad (5\text{-}3)$$

With: $similarity$ Indicates how much do two polygons overlap, considering size, shape and position in a cartesian space.

          $area(P)$ Area of a polygon P

The concept of "similarity" described in (5-3) is applied to compare the FOR polygons obtained from both methods. The function compares the shape, size and position of two polygons in the complex power domain. A 100% similarity means that the two polygons have the exact same shape and are located at the same position in the complex power domain. On the other side, a 0% similarity indicates that both polygons do not overlap at any point, however, even if the polygons have the exact same shape.

The CIGRE MV grid model of Figure 5-1 is adapted for this analysis. All loads and generators at buses 4-12 are replaced with FPG. The initial operation point of each FPG is defined as $0 + j0\ MVA$. For each FPG, a capability chart is assigned, following two methodologies. In the first one, rectangular capability charts with limits at ±1 MW/MVAr and ±1.5 MW/MVAr are defined for all FPG equally. In the second approach, the FPG are assigned a random convex polygon with 4 to 12 vertices each. The location of the randomly selected vertices is confined within a box defined by ±1 MW/MVAr and ±1.5 MW/MVAr. All polygons contain the origin of the complex power flow cartesian space within their boundaries. The same defined FPG capability charts are used in both FOR computation methods (LFA and RS), in order to allow a comparison between them. In Figure 5-21, an example of the randomly selected polygons following the aforementioned prescriptions is shown. These polygons are used for the subsequent case study.

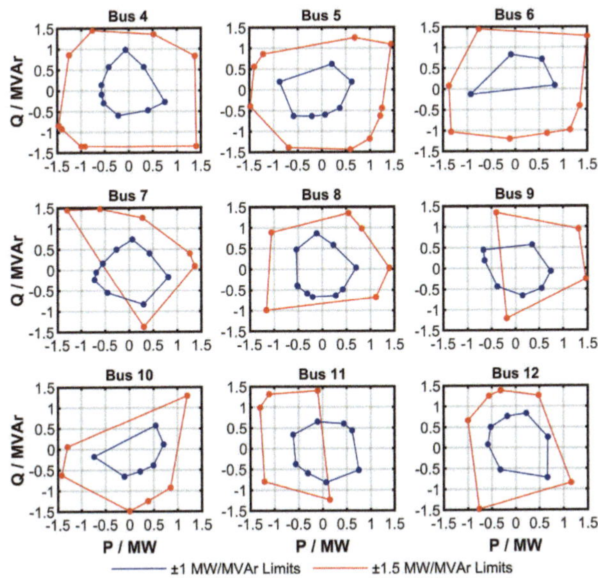

Figure 5-21: FPG random capability charts connected at buses 4-12. [108]

A relevant aspect is that the permutations of the vertices of the polygon combinations (assuming between 4 and 12 vertices each) would yield something between 10^6 - 10^{11} different IPF, and would require as many power flow calculations. The RS approach intends to reduce the number of calculations, by assessing just a reduced number of random samples. Initially, the FOR of the defined grid is computed using 10^5 random operation points bounded by the rectangular FPG constraints (±1 MW/MVAr and ±1.5 MW/MVAr). The following PDF are used in this analysis: Normal, Uniform, Beta and Rademacher [108]. Results are compared to the LFA method. The results, compared to the LFA method, are shown in Figure 5-22 for the ±1 MW/MVAr case and Figure 5-23 for the ±1.5 MW/MVAr case. The analysis is performed in both cases with and without considering grid constraints.

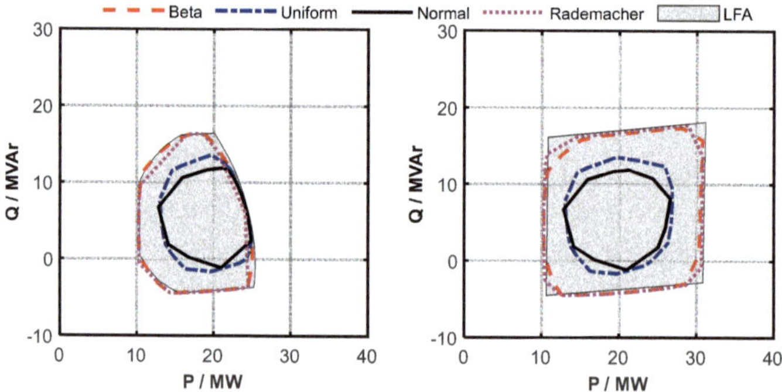

Figure 5-22: Comparison between RS and LFA methods with square FPU boundaries restricted to ±1 MW/MVAr limits. (left) With grid restrictions, (right) Without grid restrictions.

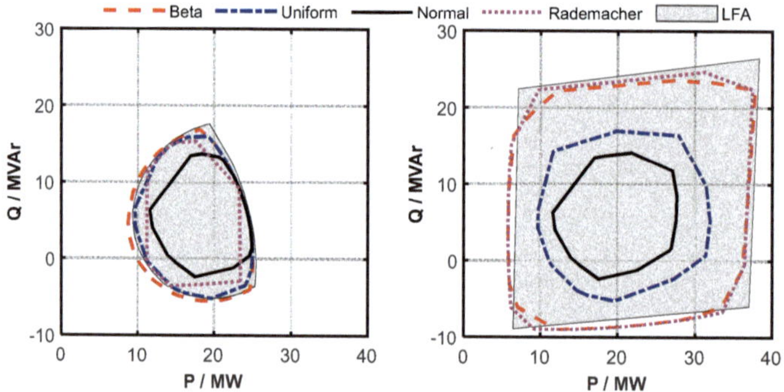

Figure 5-23: Comparison between RS and LFA methods with square FPU boundaries restricted to ±1.5 MW/MVAr limits. (left) With grid restrictions, (right) Without grid restrictions.

The Rademacher PDF behaves similarly to the beta PDF, results that resemble the shape of the FOR obtained using the LFA method. On the other hand, the normal and uniform PDF clearly underestimate the size of the FOR, with conservative results. Both Rademacher and beta PDF help improving the quality of the RS method, proving to be a valid solution to the concerns raised in [23], while corroborating the reasoning provided in [7] and [14]. This demonstration is initially valid just for the case of a few (nine in this case) rectangular FPU limits.

One challenge is to replicate the experiment considering non-rectangular convex polygons as the FPG boundaries. The problem lies in how to produce a PDF that resembles the Beta and Rademacher PDF, which can be used with generic convex polygons. The proposed methods to generate bivariate PDF were introduced in the previous chapters, and their usage with the polygons of Figure 5-21 are shown in Figure 5-24 and Figure 5-25, once again compared to the LFA method. The analysis considers 10.000 random samples using each PDF. The following bivariate PDF were applied to the RS approach: Uniform, 2D-Histogram (Bivariate), Vertices combinations (Vertices) and Quadrants partition (Quadrants).

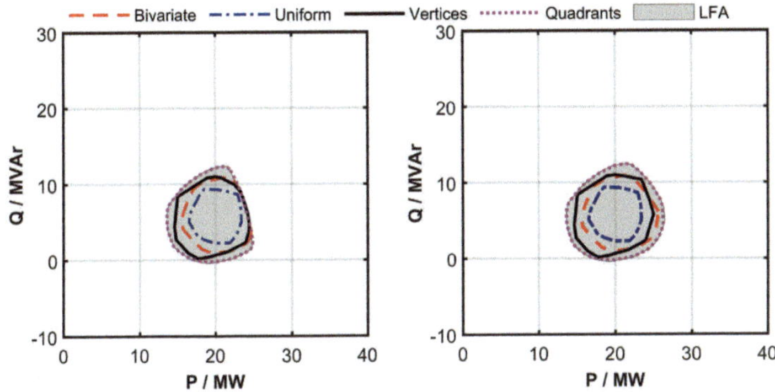

Figure 5-24: Comparison between RS and LFA methods with FPU boundaries restricted to ±1 MW/MVAr limits. (left) With grid restrictions, (right) Without grid restrictions.

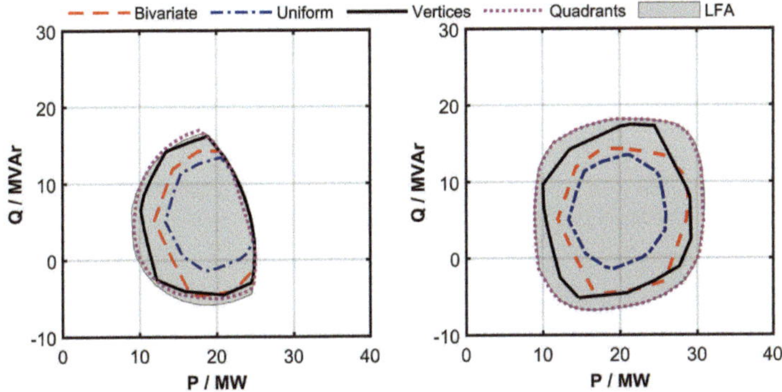

Figure 5-25: Comparison between RS and LFA methods with FPU boundaries restricted to ±1.5 MW/MVAr limits. (left) With grid restrictions, (right) Without grid restrictions.

The Quadrants PDF shows good similarity to the LFA approach, displaying its value in this simplified scenario, while the Uniform PDF shows the worst performance. The Bivariate approach shows improvements compared to the Uniform PDF, but not as good as would be expected from the Beta PDF in Figure 5-22 and Figure 5-23 for the rectangular cases. Due to the strong performance of the Quadrants approach, as well as its simple implementation, it was decided not to include the Bivariate approach in the forthcoming analysis. The grid constraints do have an impact on the results, as the thermal limit of the line connecting buses 2 and 3, as well as the transformer limitations, restrict the shape of the FOR.

To analyze the replicability of the proposed PDF, the process was repeated 200 times. For each iteration, a new set of polygonal FPG boundaries is generated, considering the ±1 MVA/MVAr and ±1.5 MW/MVAr limits. Figure 5-26 and Figure 5-27 show the distribution of the similarity for the four proposed PDF. The boxplots show the absolute max/min, the 25/75 percentiles and the average values. It can be observed that the Quadrants approach provides the best comparison to the LFA method. From these results, two conclusions can be drawn. The first one being that the proposed LFA method provides plausible results for the FOR, that are corroborated with power flow calculations. The second conclusion is that the RS approach can also provide accurate results, but only if the proper PDF is applied for the selection of the random samples. These conclusions are valid for the presented case study. This analysis was extended to a larger 20 kV distribution grid in [108], validating the tendencies observed in Figure 5-26 and Figure 5-27.

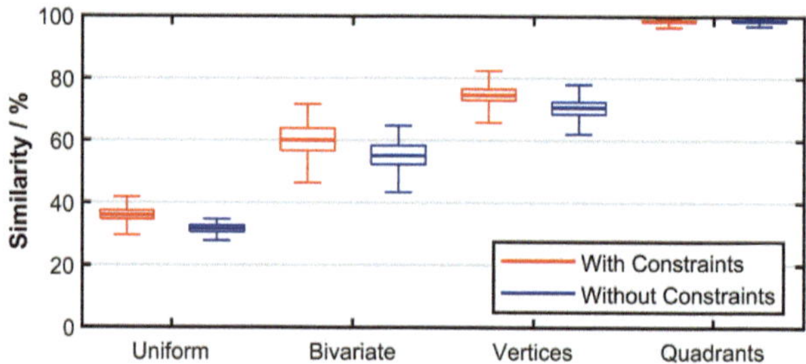

Figure 5-26: Distribution of similarity between RS approach with proposed PDF and LFA with ±1 MW/MVAr FPG limits after 200 iterations. [108]

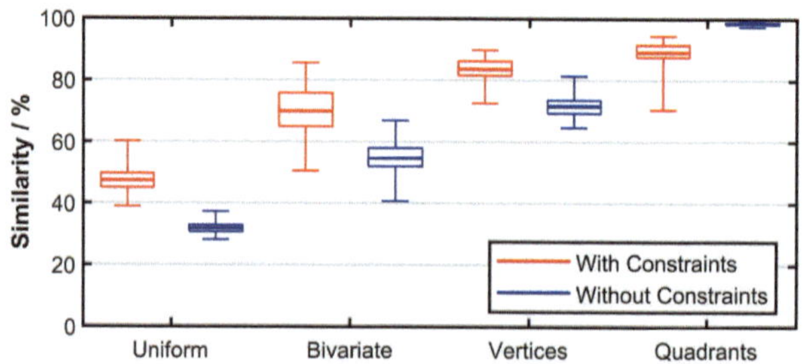

Figure 5-27: Distribution of similarity between RS approach with proposed PDF and LFA with ±1.5 MW/MVAr FPG limits after 200 iterations. [108]

5.7 Comparison of LFA to Similar Approaches

The LFA algorithm shares some similarities to the ICPF algorithm of [22], against which a comparison was done in [4], regarding the quality of the FOR and the computation time. This showed the strengths of the algorithm proposed in this thesis. Meanwhile, further developments to the LFA were done, allowing for a greater reduction of the computational time, i.e. parallel computing techniques in the quadrants-based search of FOR boundary points and using the commercial CPLEX solver instead of the *linprog* function in MATLAB. This chapter provides a comparison of the LFA, RS and ICPF algorithms applied to the scenario described in [108].

The radial MV Feeder 4 (Figure 5-4) is used in [108] to study the limitations of using the RS approach based on the amount of FPU/FPG in a grid. The analysis complements the results of [4]. The approach provided in [108] is used in here, where the FOR is computed with different numbers of FPG, each with a random convex capability chart. The sizes of the random capability charts are bounded to ±200 kW and ±200 kVAr, similar to Chapter 5.6. All polygons contain the origin of the complex power flow cartesian space within their boundaries. Figure 5-21 acts as orientation for the approach taken to define these polygons. In each iteration, equal to increasing the number of FPG in the grid (between 1 and 95 FPG), the FOR is computed using the RS, ICPF and LFA algorithms. The entire process is repeated 30 times in total, in order to have a statistical support of the results.

The ICPF method was implemented here with the non-linear equations system and adapted with the improved search algorithm proposed for the LFA algorithm (cf. Chapter 4.2.9, equations (3-17) - (3-20), and constraint (4-10) [22]). The *fmincon* function in MATLAB is used to solve the non-linear OPF to locate the FOR vertices, while the dual-simplex CPLEX solver is used in the LFA algorithm. The computational speed of both, ICPF and LFA, methods was improved by parallelizing the computation of the four quadrants. Additionally, the RS approach is implemented with a linear power flow algorithm [145]. This allows for much better computational times than performing sequential NR-PF calculations. The maximal amount of FOR points searched by both ICPF and LFA methods is limited to 32 ($k_{max} = 3$). The RS approach considers 250.000 samples, obtained using the quadrants PDF described in Chapter 3.2.3.2. The similarity of the FOR obtained with the different methods, as well as the required computational times for the selected scenario are analyzed next.

More FPG means that more flexibility can be used in the grid, resulting in a larger FOR; until grid constraints begin restricting the area. Figure 5-28 shows how the FOR size grows with increasing numbers of FPG. It was shown in [108], that the

RS method with the Quadrants approach allows the size of the FOR to growth with the number of FPG, thus the quality decays quite rapidly compared to the LFA method (this becomes more clear in Figure 5-28 and Figure 5-29). In contrast, LFA shows a very good similarity compared to the adapted ICPF method, where the differences between both FOR are almost neglectable in most cases.

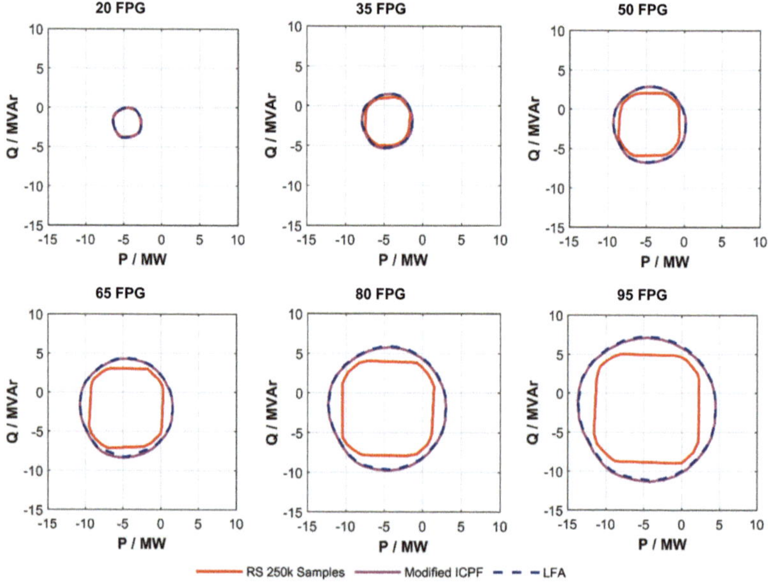

Figure 5-28: FOR obtained with RS, ICPF and LFA methods with increasing number of FPG.

The similarity is computed with (5-3), and the FOR obtained with the LFA is used as base for comparisons. Both algorithms provide similar results, yet the LFA computes faster (Figure 5-30)[14]. The processing time of the RS method is independent of the number of FPG, while the computation time increases linearly in both ICPF and LFA methods. As the number of FPG increases, the non-linear method becomes more unstable, and the processing time increases greatly.

Different factors impact the computation time of the LFA method, e.g. selected optimization solver, or the parametrization of the algorithm itself. The most important parameter in the LFA algorithm is the maximal amount of points that can be assessed during one aggregation. This results in at most $4 + 2^{k_{max}}$ FOR boundary points, where k_{max} is the number of iterations as previously defined in Figure 4-4. With the selection of a larger k_{max}, the required computational time

---

[14] Simulations performed in Matlab 2020b using a Ryzen 7 3700X at 3.6 GHz and 32GB RAM.

increases, as can be clearly observed in Figure 5-31. Increasing the number of points improves the quality of the FOR, at the cost of drastically increasing the computational time. Therefore, a trade-off between time and quality needs to be reached. A good compromise is achieved between $2 \leq k_{max} \leq 4$ iterations (resulting in 16 to 64 FOR boundary points), depending on the size of the problem.

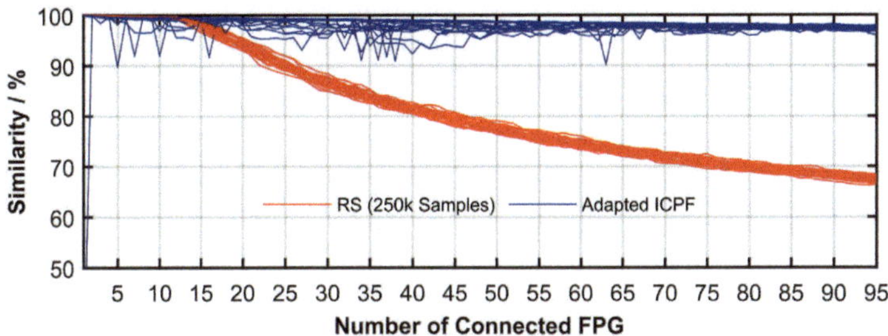

Figure 5-29: *Similarity of RS and adapted ICPF approaches compared to LFA with increasing number of FPG.*

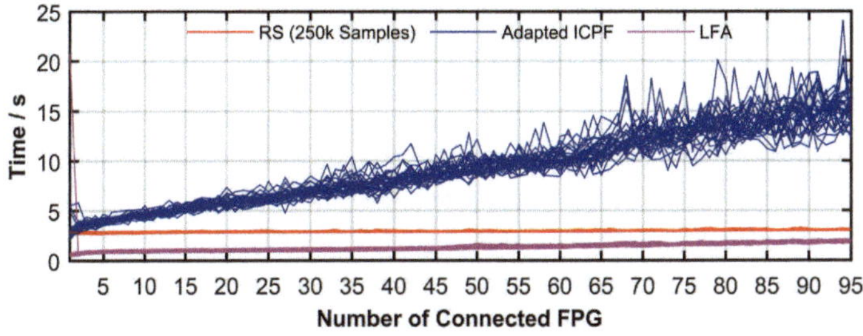

Figure 5-30: *Computation time with RS, adapted ICPF and LFA methods based on FPG number.*

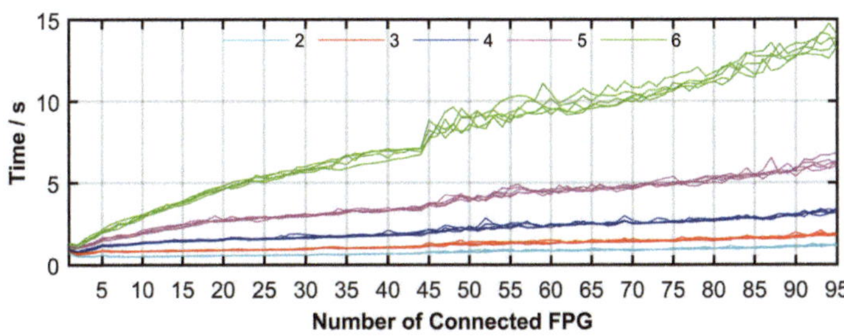

Figure 5-31: *Required computation time to obtain the FOR using the LFA method with increasing number of iterations k_{max} to assess additional boundary points.*

6   Use Cases for the Planning and Operation of Distribution Grids

In the previous chapters, a method to compute the FOR of distribution grids based on linear optimization was introduced and compared to similar methods. The main feature of the proposed method is that it strongly reduces the required computation time to assess a FOR. Different use cases for the use of the FOR in the planning and operation of power systems can be found in literature; however, there is a wide consensus that a reduction in the computational time would be beneficial for the further development of the use cases. By reducing the computational time, the FOR concept can be projected to be used during "real-time" operation of power systems, which relies on fast decision in response to the current system state. The application of the LFA method would allow to reengineer the usage of the FOR in existing use cases, which were forced in many cases to consider either simplified solutions (random sampling approach) or to keep its usage restricted to reduced numbers of FPU/FPG.

Some use cases have been published linking the FOR to the operation of LV microgrids. In [113], the COMMELEC framework showed the use of the FOR concept in the real-time control of a LV microgrid, with a dazing refresh rate of 200ms. To meet requirements of real-time applications, grid losses are neglected in the FOR-computation process, which relies on Minkowski sums. Yet, it is unclear how the FPU are modelled, which is a relevant information necessary to estimate the computation time of the FOR. Therefore, the scalability of the approach for larger LV grids or even MV grids remains uncertain. A similar concept was presented in [148], which uses the FOR of a microgrid to determine proper droop control gains parametrization at the PCC, which is then distributed to the local controllers of the DER. The hierarchical approach resembles the schemes proposed in [41] and [149], and the application to microgrid stability control approaches looks promising. Nevertheless, the publication is simulation-based and there is no further information about the "real-time" capability of the approach. In [150], the FOR is used to define the feasible operation points that can be defined during the resynchronization of an islanded microgrid.

Different usages of the FOR during the daily grid operation by TSO/DSO have been proposed as well. Throughout the IDEAL project, a system was developed to demonstrate how a distributed coordination of power flow control actions can help relieving congestions in distribution lines [57]. The FOR is applied to simplify the representation of underlayered grid sections, which helps the grid operators in the control room to optimize the reactive power flow by adapting the line impedances through Distributed Power Flow Controllers (DPFC). Such an application, aimed to improve the observability of the grid, is also found in [39], which took place under the Italian ATLANTIDE project. In [151], the use of the FOR concept to deal with grids with limited observability was discussed. In it, a distributed hierarchical approach uses the FOR of underlayered grids to assess possible operation point changes that could help remedial N-1 security violations at the higher voltage level.

Not only grid operators might benefit from the FOR concept applied to power systems, but also market participants. A model for the usage of the FOR concept to the operation and marketing of a VPP was provided in [152]. There, the FOR computation is performed solely for the assets included in the VPP, allowing to describe their flexibility potential over time. A novel aspect in that work is that the FOR computation is included within a techno-economical optimization, which aims to maximize the profit of a VPP within a market environment.

There has been plenty of activity around the use of the FOR in the last few years, not only in the grid planning aspect, but also in the grid operation area. The goal of this chapter is to provide insight of how a fast computation of the FOR can provide additional benefits to the discussed use cases and even help to generate new use cases. This is discussed based on three different aspects, which show some resemblance to the ones portrayed before. The first use case involves the coupling of the algorithm with time-series, which usually requires the sequential computation of a large chain of FOR. The second use case corresponds to a multi-level aggregation approach, which aims to distribute and parallelize the computation of distribution grids over different voltage levels. The third, and final, use case aims to using the resulting FOR in the provision of ancillary services as required by an overlayered grid operator in a hierarchical operation. As mentioned before, these use cases have been partially discussed in other publications, as well as by the author of this thesis (e.g. [41], [141], [150]). The novelty in this instance is the application of the fast calculation approach for more complex analysis of power grids based on a large number of repetitions of the computations.

6.1 Time-Series Based Aggregation

A recurrent task in the planning and operation of distribution grids involves working with prognoses of load and generation, in order to assess future scenarios of the grid. These can be either short-timed (i.e. the next 15 minutes or the coming few hours) or long-termed (i.e. a month or an entire year). Both of these timeframes are covered in this chapter.

6.1.1 Long-Term Time-Series Analysis

As RES penetration introduces new uncertainties in the power systems, traditional worst-case grid planning approaches are being replaced with time-series based analysis (e.g. [153]). Time-series based calculations allow assessing several aspects that are not possible in a worst-case approach, i.e. the simultaneity of load and generation, seasonal behaviors of the grid, or the probability of overloading certain power lines or transformers. In general, such methods provide information of better quality for grid reinforcement planning purposes, compared to traditional worst-case approaches. Coupling the FOR with time-series can provide even more additional information, as for each period, not only a single operation point of the grid is obtained, but also the possible operation points that could be achieved when controlling under layered FPU.

The integration of time-series into the FOR concept is demonstrated using the MV grid of Figure 5-5 as reference. Loads are defined as the aggregated residual load connected to the LV/MV transformers, for which time-series in 15-minutes resolution for an entire year are generated (35040 periods). The time-series are a combination of real measurement data (for large customers and generators) and synthetic profiles (based on [154], for buses without measurement). These time-series are coupled to the FPU following the approach introduced in Chapter 4.5. As no information regarding reactive power consumption was available, a random power factor is defined for each LV grid, with the following normal distribution:

$$cos(\emptyset)_{ind} = \mathcal{N}(\mu, \sigma^2) = \mathcal{N}(0.95, 0.025) \qquad (6\text{-}1)$$

With: $\mathcal{N}(\mu, \sigma^2)$ Normal distribution with mean μ and variance σ^2

This example aims to show, how the flexibility provision of PV generators impacts the possible IPFs of the grid, observed at the HV/MV transformer. The results of the time-series-based FOR computation of four different months is shown in Figure 6-1, in order to show the seasonal impact of the PV generators in the FOR (all PV are defined as Type 5 FPU). Two scenarios were computed, one with the original dimensions of PV (blue area) and a second one with 200% PV capacity (red area). In this case, the grid restrictions play no role in the results, as the power lines and transformer are never overloaded, at the same time the voltage limits are always respected. Increasing generation in this particular grid is actually beneficial, as it helps reducing the peaks of the load profile.

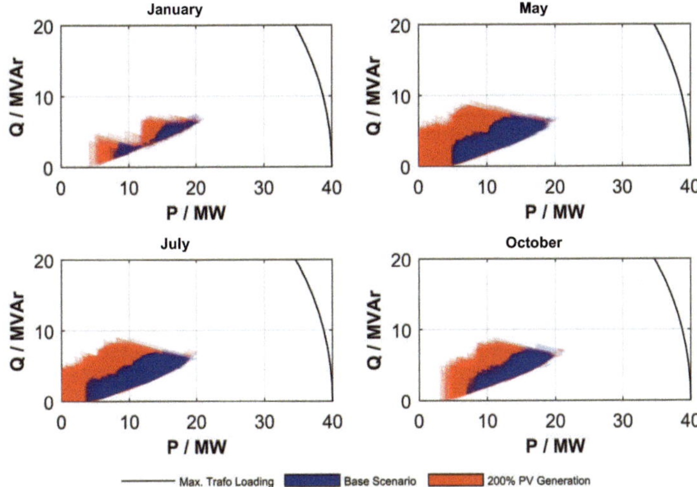

Figure 6-1:   Seasonal impact of PV generation in the FOR (PV increased in 200%).

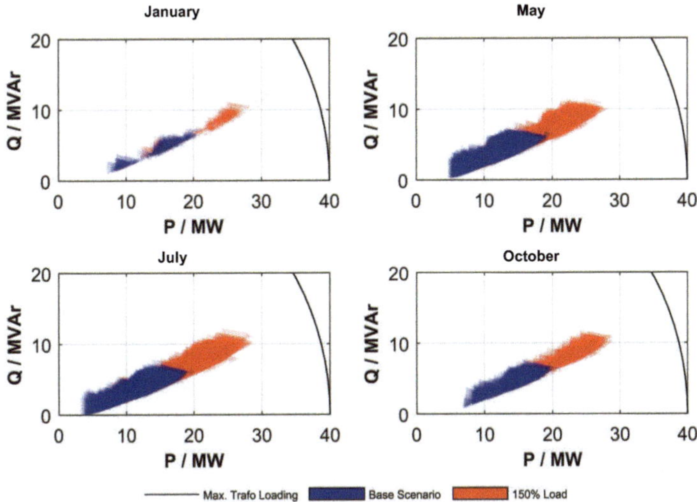

Figure 6-2: Seasonal impact of PV generation in the FOR (load increased in 150%).

On the opposite, when the load is increased, to 150% of the nominal value in this case, the IPF are shifted closer to the limits of the HV/MV transformer, as can be seen in Figure 6-2. The resulting FOR are a combination of the results that would be obtained after applying worst-case and probabilistic approaches to assess the future status of a given power grid. This provides more information than a worst-case approach by showing how IPFs behave over time, e.g. through the overlapping of the areas in the graphic, which in turn is equivalent to the main result of a probabilistic approach. As is visible from Figure 6-1 and Figure 6-2, the selected grid has enough room for increasing load and generation, at least at a global level, a more detailed analysis of the single branches is still required. However, this is could be assessed with help of this method, as shown in the following example. By assuming reduced branch limits (50% of the nominal value), the grid capacity is artificially decreased. On one side, only the transformer is limited, followed by a reduction of all branch capacities (Figure 6-3). For the sake of demonstration, the month of May is used as reference, considering an increase in the load of 125%. Reducing the transformer capacity would already make the operation of the grid unfeasible during plenty of periods. This becomes worse by halving the entire grid capacity, where the blue area becomes much more constrained compared to the base case, meaning that the internal branches are also overloaded. Similar analysis can be performed to assess bus voltage violations; which is not an issue in this specific scenario.

Additionally, the provision of flexibility by other means of generation connected to the grid (i.e. CHP, Biogas or Hydropower) can be analyzed following similar approaches. Different to RES-based generation, the flexibility provision of SG is not directly associated to climate factors, therefore, their flexibility provision capability depends on how they are dispatched. In the analyzed grid, one diesel generator with 640 kW of installed capacity is connected (FPU of Type 6). The computation is repeated, resulting in a generally larger FOR through all periods, because of the SG capability (a control over the entire feasible region is assumed), as seen in Figure 6-4. In normal operation, conventional generators would more likely be acting as base load power plants, however, if they are large enough, they could be either directly marketized for some ancillary services or integrated into a commercial VPP concept. In the later cases, it makes sense to consider possible changes to their operation point as part of the flexibility analysis. The flexibility provision of SG can be seen as time-invariant, making the FOR appear as time-invariant as well, regardless of the time of the year.

Assessing the FOR of an entire month with 31 days requires 2976 computations in total (15-minutes intervals). This could be performed sequentially, however, the large amount of required iterations demands large processing times, which can be shortened by parallelizing the FOR computations. This is done using the parallel computation capabilities of MATLAB, considering that each FOR computation is independent. A comparison of the total computation times with and without parallel processing is shown in Figure 6-5. A reduction of around 90% of the processing time can be achieved with little effort.

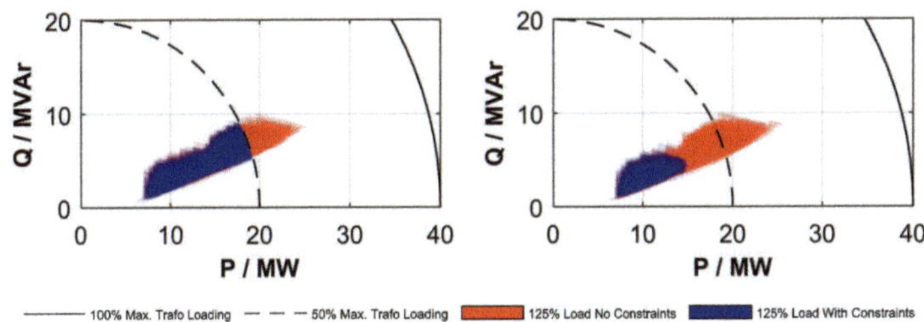

Figure 6-3: Impact of branches constraints in the FOR. (left) Only the transformer is limited to 50% of the original capacity, (right) All branches are limited to 50% of their original capacity.

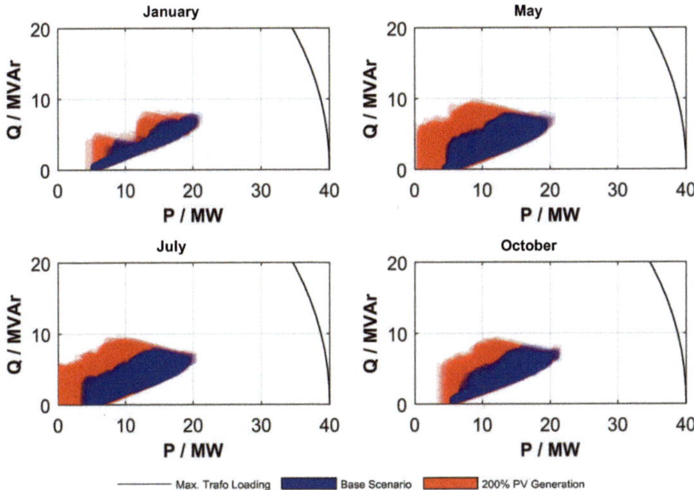

Figure 6-4: Seasonal impact of PV generation in the FOR (PV increased in 200%), considering time-invariant flexibility provided by a synchronous generator.

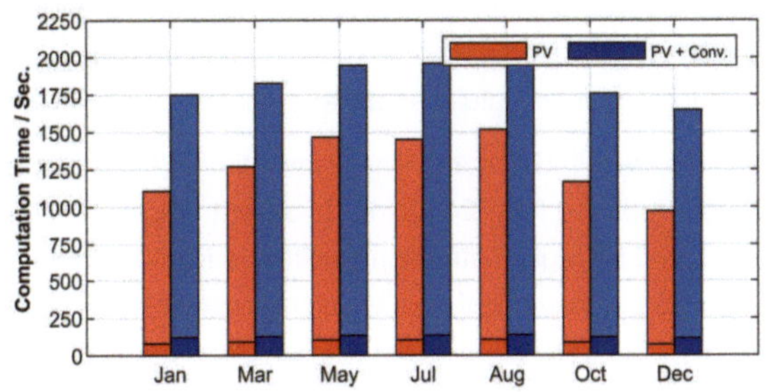

Figure 6-5: Computation time to assess the FOR of a MV feeder over an entire month, serial computation in lighter color, and parallel computation in darker colors.

The processing time of the LFA method was already analyzed in Chapter 5.7, nonetheless, those results are complemented in Table 6-1, which shows the seasonal variation of the average FOR computation time. It can be seen how the average values correlate with the PV generation over the year, taking less time to compute in winter and longer in summer, where the days are longer, hence PV generation lasts longer. Adding the SG generator increases the overall computation time, as it can provide flexibility the entire time.

Table 6-1: Summary of average FOR computation times in different scenarios.

Scenario (Month)	PV / s	PV+SG / s
January	0.35	0.55
March	0.40	0.57
May	0.46	0.61
July	0.45	0.61
August	0.47	0.62
October	0.36	0.55
December	0.30	0.52
Year Average	0.40	0.58

6.1.2 Short-Term Time-Series-based Analysis

The operation of power systems requires making short-term decisions, therefore, complex calculations must be performed in reduced time, in order be able to ensure a correct operation of the grid in "real-time" and allow hasty curative actions when an asset fails or there is a deviation from the expected operation. Several mechanisms have been developed for this purpose, i.e. frequency control, reactive power compensation, or redispatch, which have allowed to fulfill the tasks successfully for many years. The integration of volatile RES, storage systems, and controllable loads, in conjunction with the decommission of conventional power plants, are forcing successfully established processes to be modernized.

One example can be seen in Germany, where the Primary Control Reserve (PCR) auction blocks were reduced from 24h to 4h, in order to facilitate the participation of wind generation and storage systems in the frequency control ancillary service market[15]. This signalizes changing times, which harmonizes with current research trends focusing on TSO-DSO cooperation, on defining novel flexibility markets, and generally taking advantage of increasing RES controllability and storage pos-

---

[15] www.regelleistung.de

sibilities [155]. Another example is the *Redispatch 2.0* being worked on by German TSO and DSO, aiming to include RES and CHP systems of 100 kW or more as well as systems with a lower output that can be remotely controlled by a grid operator at any time in the redispatch process [16].

In this context, the development of modern tools to assess flexibility provision and its impact on the grid becomes necessary, especially considering the countless different types of ancillary services available. Ancillary services are per definition short-noticed actions, which result from day-ahead, intra-day or real-time decisions, with the tendency to move closer to real-time, meaning that the information flows need to be fast. A fast computation of the FOR can help in this matter, as it allows to provide complex input to the grid operator in reduced time, matching the aforementioned requirements. In this use case, a short-time grid behavior forecast is proposed, in which possible operation states of a grid segment are analyzed considering flexibility capabilities. This is based on short-term time-series, which forecast the behavior of grid assets (load and generation, particularly crucial for RES) for the next instants, i.e. next 15 minutes, next 4 hours, or even the entire next day. By computing the FOR, it is possible to observe, if the potential use of specific ancillary services (e.g. PCR) can impact the normal operation of the underlayered grid, as well as how much flexibility can be provided in this interval or if it is possible to sustain specific operation points at the TSO/DSO interface over time. This can be performed easily in grids with few controllable assets, however, the complexity increases with the growing number of installed DER.

The MV feeder of Figure 5-5 is considered for this example, mostly due to its large installed PV power, with 6.9 MW in total. Based on the data visualization described in [141], the aggregation over an entire day is done. Figure 6-6 (left) shows the resulting FOR and the corresponding IPF for each interval, while in Figure 6-6 (right), the initial IPF is subtracted from the FOR. This shows the flexibility that can be provided for each period regardless of the grid operation point. From this perspective it can be seen in which direction (regarding active power) flexibility can be provided. In this case a negative ΔP is obtained by curtailing PV, while a positive ΔP comes from activating SG.

Some clusters of conceivable IPFs can be observed in Figure 6-6 (left) where the FOR overlap. This visualization for an entire day does not offer much information regarding the possible IPFs that can be achieved, as there are clear differences between the night and day behaviors. Focusing on just a few periods make more

---

[16] Officially beginning on 01.10.2021, www.bdew.de

sense, especially considering the typical time-frames for ancillary services provision, e.g. 4 hours for frequency-based ancillary services.

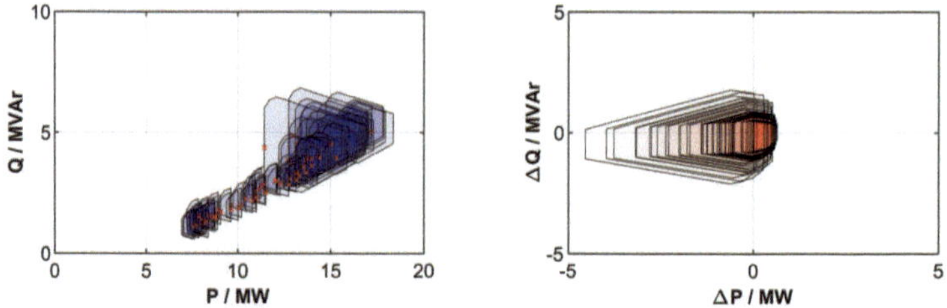

Figure 6-6: Evolution of the FOR during an entire day (96 periods). (left) FOR including initial IPF, (right) FOR minus the IPF.

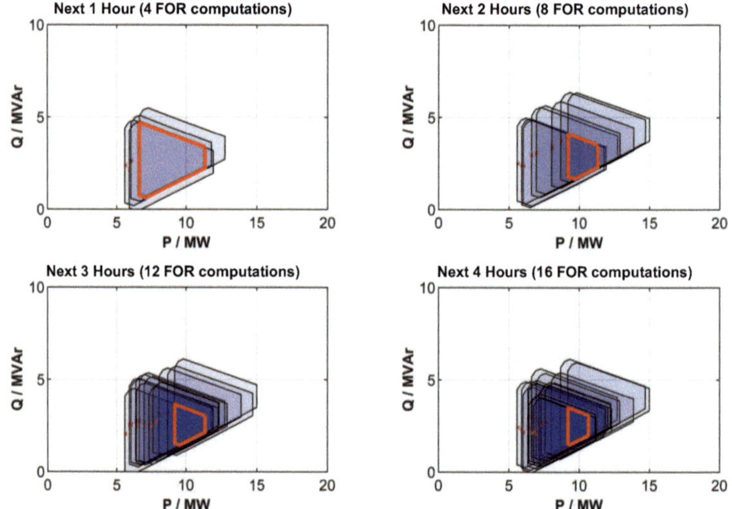

Figure 6-7: Sequential FOR computation considering different time spans. The $FOR_{100\%}$, i.e. the red polygon, becomes smaller as the time span increases.

What is shown in Figure 6-7 are sequential FOR in different time spans and the set of IPF that are possible to reach 100% of the time ($FOR_{100\%}$), which is the intersection of all polygons, defined as:

$$FOR_{100\%} = \bigcap_{t=t_0}^{t_0+n} FOR_t \qquad (6\text{-}2)$$

With: $FOR_{100\%}$ FOR including operation points achievable 100% of the time

The resulting FOR are shown as the shaded areas, while $FOR_{100\%}$ is represented as the red polygon. As 15-minutes time-steps are used, four consecutive FOR amount to one hour. In the four presented cases, the initial time-step t_0 is always the same, only the number of future time-steps n is changed. The further one looks into the future, the smaller is the resulting polygon, which in some cases can be even inexistent (e.g. midday ramp).

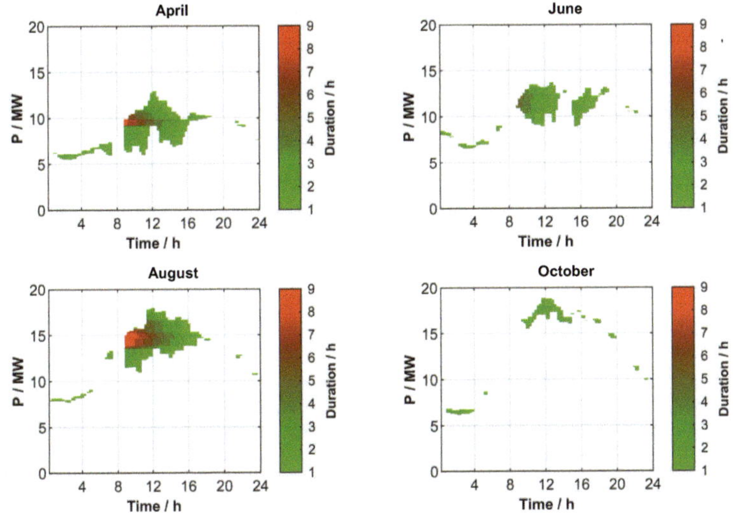

Figure 6-8: Set of IPF that can be sustained for longer time-periods. The color map describes how long an operation point can be sustained starting from that specific time.

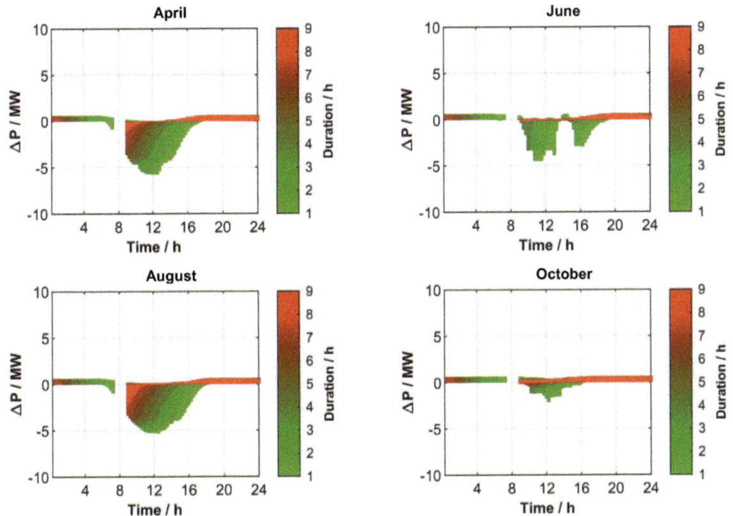

Figure 6-9: Flexibility that can be sustained for longer time-periods. The color map describes how long an operation point can be sustained starting from that specific time.

The $FOR_{100\%}$ not only depends on the size of the FPU boundaries, but also on the operation point of the grid, which fluctuates with load and generation over time. By applying the LFA method, it is possible to compute these scenarios in a short time, allowing to compute the $FOR_{100\%}$ for several sequential time-frames.

Figure 6-8 depicts the IPFs that could be maintained for at least one more hour, for four different days. The color scale shows for how long an operation point could be maintained from that specific moment onwards. It is clear that just a few IPF can be maintained for the maximal assessed time (red color, representing 9 hours in the future) is reduced. The shape and colors depend on the conditions of each grid and the amount of flexibility available. In the example, during the midday peak in a summer day in August, a constant operation point around 13-15 MW could be sustained for around 9 hours (beginning at 9AM). While in June, some operation points could be sustained only for a couple of hours. In contrast, a colder day in October, where PV generators offer only limited flexibility, shows a reduced number of IPF that can be held over time.

The example in Figure 6-8 shows the probabilistic behavior of the IPF over a time span, yet it does provide much insight into the actual amount of flexibility that the FPU inside the FPG can provide. This can be seen in Figure 6-9, which shows for the same four days how much the IPF can be modulated for each period and for how long that change can be maintained. During the day, PV generation takes over and offers the most amount of negative flexibility, as they can be curtailed. There is an emergency diesel SG connected to the grid, which can provide positive flexibility for the most part of the day. Therefore, flexibility in both directions can be offered.

A noticeable discontinuity of the color band can be observed in the early morning hours in Figure 6-8 and Figure 6-9, between 8:00 and 10:00. During this time, at sunrise, is when PV generators begin injecting power into the grid with a large ramp. This causes a state where the flexibility provision from the grid cannot be sustained for longer than one hour, and those FOR are neglected in the diagrams.

6.2　Multi-Level Aggregation

Power systems are incredibly complex, with thousands or even millions of assets spread over large distances, that require constant surveillance and precise control, with a constant risk of failure. Due to its complexity and size, the grids are supervised and controlled by different entities and control systems, depending on the operating voltage level, as well as geographical, geopolitical and economical restrictions, among others. Germany is an extreme case, with 4 TSO and close to

900 DSO of vastly different sizes operating the power grid. The level of coordination, among these companies, that is necessary to ensure a stable operation of the power system is enormous, which in many cases is hindered by the fact that they are not allowed (or in some cases just unwilling) to share detailed information about the topology of their grids. This would in many cases facilitate the coordination efforts among grid operators sharing physical interconnections.

This chapter discusses a method that would enable information exchange among grid operators, by using the FOR to communicate the state of the grid. This would not only allow improving the communication between grid operators by reducing the amount of information required to be transferred (just the FOR), but also by reducing the complexity of the computation of the FOR. The FOR concept allows for representing entire grid sections through simple models, i.e. a vector with PQ coordinates, which can then be used as input for the assessment of the rest of the interconnected networks. With the use of aggregation methods like the LFA, this entire process can be performed in short time, and can be reduced even more by parallelizing the computation of the single underlayered grids.

A concept for the distributed computation of the FOR, focusing on vertical grid coordination, was proposed in [55], where the FOR is computed in a bottom-up manner among different voltage levels. Parallelly, a similar approach was taken in [41], where the operation of microgrids was analyzed by means of the FOR computation. Figure 6-10 shows how the operation of a grid can be divided according to the different voltage levels (e.g. single MV or LV feeders), as in reality they are all part of one larger and more complex system. With the FOR concept, the operation of the underlayered grids can be represented as equivalent generators [39].

Following the structure of Figure 6-10, in a multi-level aggregation approach (i.e. bottom-up aggregation), the FOR of the different grid sections are sequentially computed beginning with the lowest voltage grid levels (e.g. LV grids). A similar iterative approach is proposed in [156], although just focused on reactive power provision at the interconnection point, as the FOR concept was not considered. The resulting FOR of each underlayered grid is used to define the FPG boundaries at the interconnection point with the higher voltage level. Once the FOR of each underlayered FPG is assessed, the FOR of the next voltage level is computed. This is repeated until the highest voltage level is achieved. It is assumed in this case, that the FOR is computed referencing the buses, where the transformers are located (the interconnection buses). The next chapters show how the multi-level approach is applied to two different grids, one involving a LV/MV interconnection and a second involving a MV/HV interconnection.

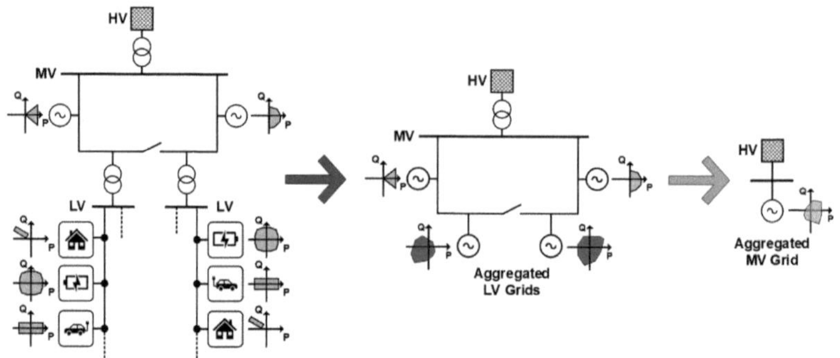

Figure 6-10: Multi-level aggregation approach, starting by aggregating the LV grids and using the resulting FOR as input for the aggregation of the MV grid. The operation of an entire MV can be described by a single FOR.

6.2.1 Multi-Level Aggregation in MV/LV Grids

In [41], a multi-stage aggregation procedure was proposed and demonstrated using the grid model shown in Figure 5-1, including the LV Feeders of Figure 5-2. There, the general functionality of the multi-level approach was demonstrated, by performing independent FOR computations of the LV grids and applying it to the MV grid. The results are compared to the equivalent computation of the FOR in a single step, where the FOR of the entire grid is assessed at once. The main result is that both FOR obtained from the single- and multi-step approaches are different, because the voltages at the interconnection points are not properly treated. A detailed description of the multi-level approach is described next.

6.2.1.1 Aggregation of Single LV Feeders

In the first step of the approach, each LV feeder is considered as an individual and isolated grid, whilst the connection to the MV grid is modelled as the slack bus. The IPF at the 20/0.4 kV transformers is aggregated considering the flexibility provision of all FPU within each of the LV feeders. The operation of each feeder can be represented through a unique FOR (for a specific operation point), from which an example is shown in Figure 6-11 for the three LV grids. These FOR are computed assuming a slack bus voltage of 1 p.u. In the analyzed case, LV grid 2 can provide the most flexibility, mostly due to the curtailment of PV generators. On other hand, the BESS provide all three feeders with positive and negative active power flexibility.

6.2.1.2 Aggregation of MV Distribution Grids

The FOR of the LV feeders represent the operation boundaries of these grids in the MV grid, as the detailed models are simplified and represented as an FPG. For the analyzed LV feeders, all three of them operate mostly in the first quadrant, acting as loads towards the MV grid. Adding the FPG to the MV grid model allows performing the next stage of the aggregation, as the FOR is once again computed, using the same methodology. The studied MV grid contains one large wind generator which provide more flexibility than the three LV feeders combined; therefore, the resulting FOR of the MV grid is mostly defined by that FPU. The resulting FOR of the multi-level aggregation approach is shown in Figure 6-12. At the same time, the FOR is obtained by assessing the entire MV grid at once, including the physical interconnections to all three LV feeders. The comparison is shown in Figure 6-12 [41]. The application of the proposed multi-level approach provides similar results compared to the all-at-once assessment of the entire grid, however, it is noticeable that both FOR are different.

During the FOR computation of the LV feeders 1 and 3, some operation point combinations cause the voltage to drop beyond the minimal allowed voltage of 0.9 p.u. These violations cannot be properly detected during the multi-level aggregation method, since the voltage at the slack bus is kept constant with the flat-start value. This issue was not properly discussed in [41], but is complemented with the next chapter.

6.2.2 Impact of Voltage at the Interconnection Point

Sectioning a grid among the different voltage levels (see Figure 6-10) results in several isolated grid sections without any physical link between them. However, in a real grid, any alteration of the voltage profile of the MV grid should inevitably be reflected in the voltage profiles of the LV feeders. By completely isolating the grid sections, this effect is entirely neglected, as the slack bus voltage (bus representing the interconnection point between grids) is now defined as a constant value for the power flow computation. One approach to overcome this issue is to complement the uncoupling of the grids with Thévenin grid equivalents, as stated in, inter alia, [7], [55], and [69]. However, it is not trivial to compute the Thevenin equivalent of a grid, especially for larger grids. Consequently, simpler alternatives are necessary.

The FOR describes a collection of operation points, each resulting in different voltage profiles of the grid. Changing the operation point of an LV feeder changes the voltage profile of the MV grid, in turn affecting the voltage profile of other LV feeders. This coupling effect can affect the computation of the FOR.

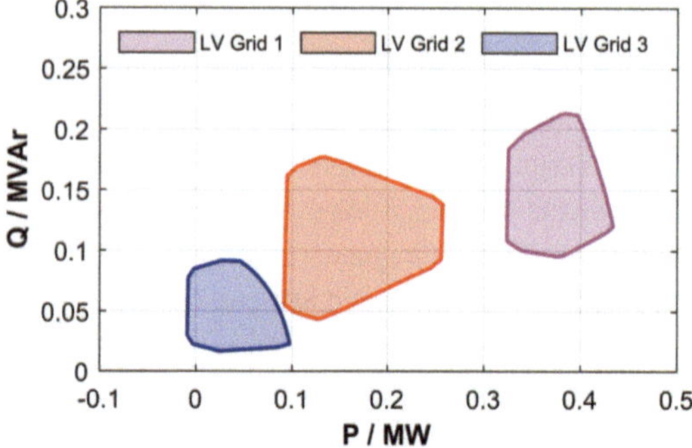

Figure 6-11: *FOR of the three analyzed LV feeders assuming 1 p.u. slack bus voltage.*

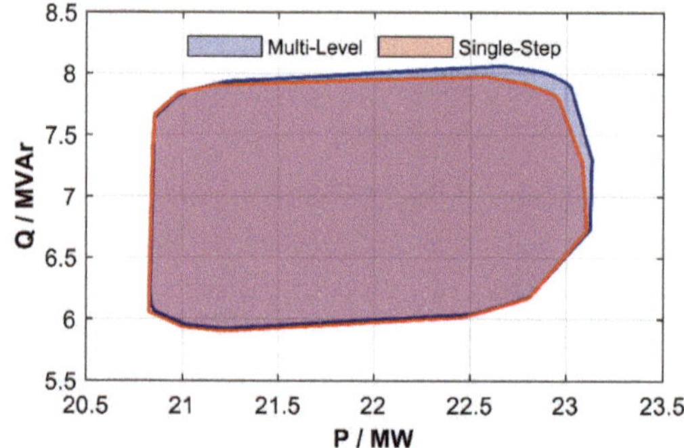

Figure 6-12: *Resulting FOR of single-step and multi-level aggregation [41].*

In [55] and [148], interesting analysis on the impact of the slack bus voltage in the FOR were provided, similar to the discussion in Chapter 5.5 of this work with the operation of an OLTC transformer. In [55], the PQu representation of a grid was proposed, corresponding to a 3D representation of the FOR, including the slack bus voltage. In the presented use case, for each of the three LV feeders, the FOR is computed for a set of discrete slack bus voltages between 0.9 p.u. and 1.1 p.u., resulting in the representations shown in Figure 6-13.

The three feeders operate in a state, where there is more load than generation, therefore, only undervoltage issues are observed. This is supported by Figure 6-13, as none of the three volumes reaches the minimum voltage of 0.9 p.u., meaning that below a certain threshold no FOR can be computed, as no valid operation points exists due to voltage constraint violations. The minimum slack bus voltage at which a valid FOR can be obtained is 0.937 p.u. for LV Feeder 1, 0.91 p.u. for LV Feeder 2, and 0.96 for LV Feeder 3. Below these voltages, the flexibility provided by the FPU in the feeders is not enough to bring the grid back to a valid operation state.

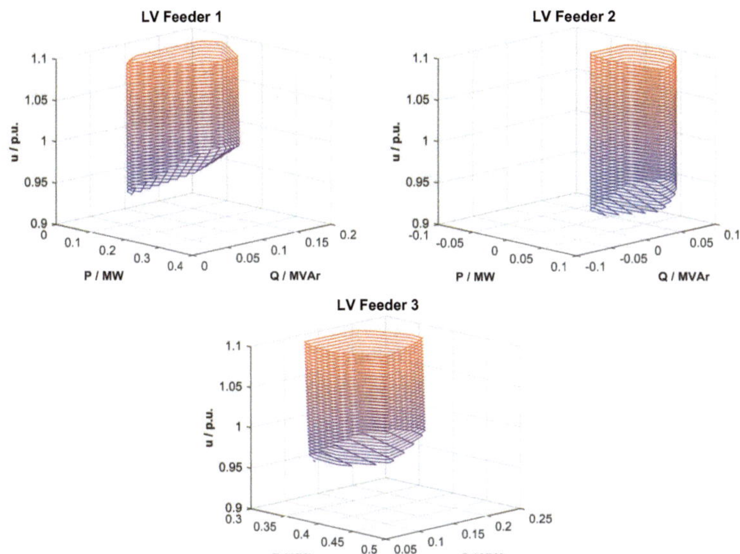

Figure 6-13: FOR of the three LV feeders computed using different slack bus voltages.

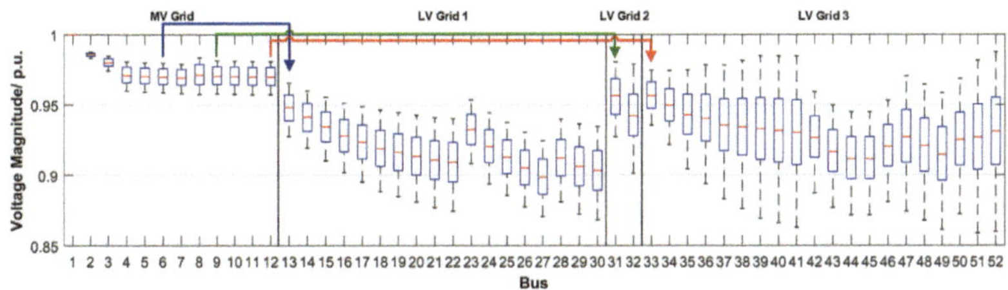

Figure 6-14: Distribution of voltages in the combined LV/MV grid for a large number of random FPU operation points using Monte-Carlo simulations[17].

It was described in [148] how the voltage profile tends to fluctuate more in a LV grid, than in a MV or HV grid. This was observed for the studied CIGRE grid with the detailed LV feeders coupled to the MV grid, creating a single large grid). Based on Monte-Carlo simulations, the impact of the FPU operation points in the grid voltage was studied.

For each FPU, a random operation point is selected, a NR-PF is computed and the resulting voltages stored, and the process is repeated 2 million times. The resulting distribution of the bus voltages is shown in the boxplot of Figure 6-14. The interconnection points of the LV feeders with the MV grid are displayed with

---

[17] Buses of LV feeders follow the same sequence as in the described models in Chapter 5.

the colored arrows. It is clear, that a small deviation in the voltage of the MV buses can cause a large deviation of the voltage at the LV level, even bringing the grid to operate below the defined limits (0.9 p.u.). When computing the FOR, this would mean that certain operation points of the LV feeders are unfeasible, confirming the results of Figure 6-13, especially as the LV feeders 1 and 3 tend to operate extremely close to the minimum voltage level.

The PQu representation can be useful when aggregating single grid sections in isolation, as it provides a good representation of the valid operation points of the grid under any slack bus voltage condition. However, it has one major disadvantage, as it requires a large number of FOR computations (because of the large number of sampled slack voltages) and the process needs to be repeated, if the FPU limits change (analogous to the time-series approach in the previous chapter). Representing the FOR with different slack bus voltages could be helpful, however, the PQu representation is not optimal and imposes additional computational burden compared to an alternative solution provided in the coming chapters.

6.2.3 Voltage Correction at Interconnection Points

The LFA method computes the FOR assuming a constant slack bus voltage, and does not consider possible changes of the slack voltage due to the usage of flexibility. The PQu concept described in [55] intends to tackle this issue, by computing the FOR of each permissible voltage at the slack bus (see Figure 6-13). A simpler approach is proposed at this point, where just a single slack bus voltage correction is performed. This allows improving the results provided by the proposed multi-level aggregation approach. The approach involves an iterative correction of the voltage at the interconnection points between grid voltage levels. A MV/LV grid is used as example for describing the method; however, the approach could be extended to other grid combinations.

The process to correct the resulting FOR in the multi-level approach:

Step 1: Prepare the grid model by setting the expected operation point of each load and generator in the MV grid and the LV feeders.

Step 2: Compute a NR-PF of each LV feeder with 1 p.u. slack bus voltage, or with a more precise value if it is already known.

Step 3: Set the resulting IPF for each LV feeder as the operation point of the aggregated LV feeder in the MV grid and compute a NR-PF.

Step 4: Set the obtained voltages for the buses connecting to the LV feeders as the slack bus voltage of each feeder and compute the FOR.

Step 5: Set the FOR as the boundary limits of the FPG representing each LV feeder in the MV grid. Compute the FOR of the MV grid.

Step 6: Each IPF in the boundary of the MV grid FOR corresponds to a specific voltage profile in the grid (analogous to the FOR-correction method of Figure 4-5). The minimal and maximal voltages for each bus are searched by analyzing these IPF.

Step 7: The FOR of the LV feeders are once again computed using the minimal or maximal bus voltage from Step 6 (this depends if the downstream lower or upper voltage limit is more likely to be violated).

Step 8: Set the FOR as the boundary limits of the FPG representing each LV feeder in the MV grid and compute the final FOR of the MV grid.

This is an iterative process, in which the voltages at the interconnection points between voltage levels are computed several times in order to replicate the coupling effect of different interconnected grid sections, where the change in the operation point of one grid section impacts the operation point of another section. The result is a FOR that is similar to what would be achieved in a single-step computation, however, an exact result (compared to the single-step FOR) is not expected to be achieved. By using the minimal and/or maximal expected voltages at the interconnection points, an underestimation of the FOR is enforced.

An example is shown in Figure 6-15, where the evolution of the FOR computed at the MV grid is shown for different slack bus voltages of the LV feeders. Based on Figure 6-13, it is known that the FOR of LV feeders becomes restricted as the slack bus voltage is decreased, an effect which is not properly assessed in cases where 1 p.u. or even the initial operation voltage is used (black and red lines in Figure 6-15, respectively). On the other hand, using the minimal voltage value (green line) is helpful to keep the final FOR similar or smaller than the single-step FOR (blue line), which would avoid showing non-valid IPF as valid.

6.2.4 Multi-Level Aggregation in HV/MV Grids

A multi-level aggregation concept was previously discussed for a MV/LV system. It was observed that in cases where the FOR is restricted by the constraints, the voltage at the slack bus of the LV feeders needs to be corrected, otherwise there is a risk of overestimating the FOR of the MV grid. In this chapter, the analysis is reproduced for a meshed HV grid serving a few MV feeders, which at the same time serve numerous LV grids. The MV feeders of Chapter 5.1.3 are defined as FPG, each supplying between 70-110 LV feeders. The remaining MV feeders are considered to be inflexible and described by a single operation point.

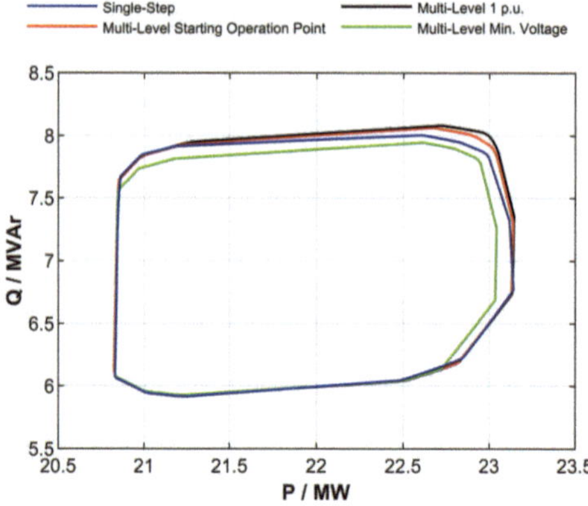

Figure 6-15: Comparison of FOR computation in single-step and multi-level aggregation with different slack bus voltages of the underlayered LV feeders.

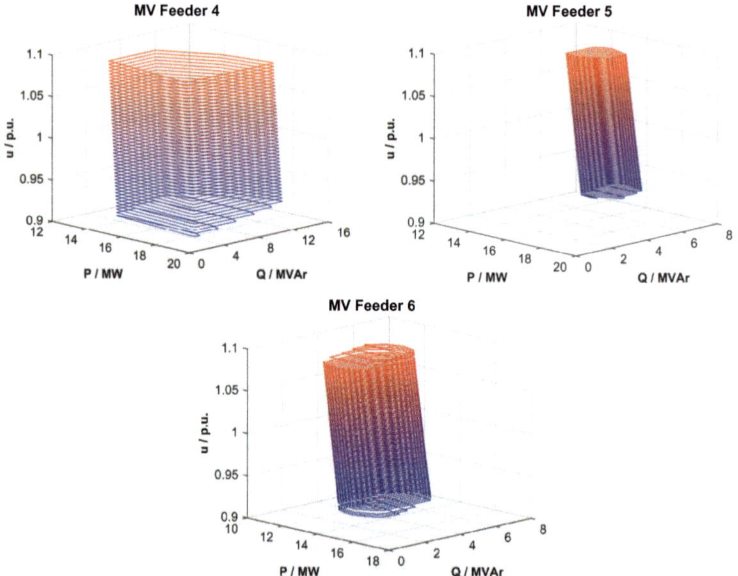

Figure 6-16: FOR of the three MV feeders computed using different slack bus voltages.

By using the PQu representation on the MV feeders, an obvious dependency of the reactive power flexibility provision on the slack bus voltage is seen. Comparing the results of the MV feeders in Figure 6-17 to the ones of the LV feeders of Figure 6-13, the volumes representing the MV feeders show a noticeable inclination to the left side, as there is a linear relationship between the reactive power limits and the slack bus voltage. The relationship is linear due to the linearity of the LFA method, otherwise, non-convexities could appear, as these MV feeders are com-posed of cables with large capacitances. This has a strong impact on the multi-level approach, making it necessary to design a proper voltage correction tech-nique. In the case of MV Feeder 6, the size of the FOR is reduced as the slack bus voltage gets closer to the limits, however, it is unlikely for a HV grid to be operated so distant from the 1 p.u. value, reducing the relevance of this issue.

The multi-level aggregation process is applied to this grid, beginning this time from the MV level. For each FPG a FOR is computed and its shape is defined by the mix of FPU within each FPG. The results are shown in Figure 6-17. It is clear that the feeders provide different amounts of flexibility. The resulting FOR of the multi-stage aggregation approach, observed at the slack bus of the HV grid, is also shown in Figure 6-17. An initial slack bus voltage of 1.04 p.u. is set for all feeders, which is corrected during the process.

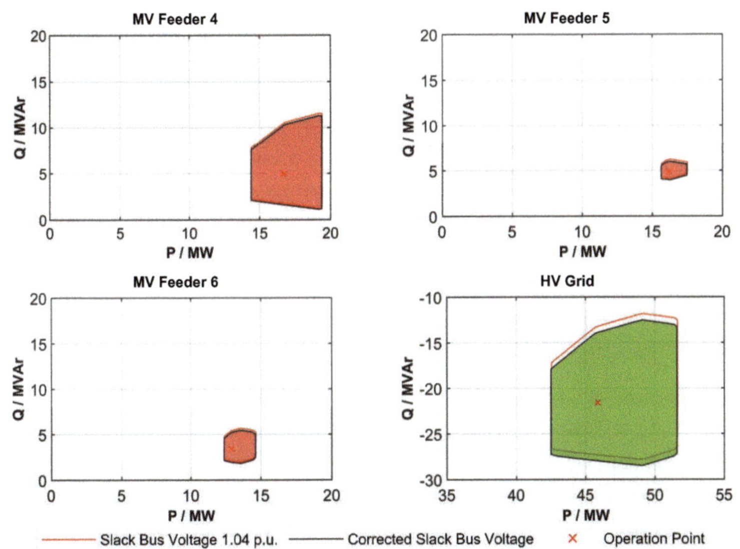

Figure 6-17: FOR with bottom-up multi-level approach and impact of slack bus voltage correction.

As previously shown in Figure 6-14, the voltage at the MV level does not fluctuate as intensely as in the case of LV grids, being also the case for HV grids. Therefore, for this example, a change in the slack bus voltage of the MV feeders does not cause a significant change in the FOR, especially as the MV grids are usually operating away from the operational limits. However, as shown before, the voltage and reactive power limits are strongly interconnected, causing the FOR to shift among the reactive power axis after the slack bus voltages are corrected.

The application of the LFA method in this scenario poses two main benefits. The first one is the quickness of the operation. Assessing the entirety of the grid with the multi-level approach required on average 18.9 seconds[18] with different operation points[19]. Table 6-2 describes the average computation times required for all stages of the process. All computations are performed serially. Overall, the grid contained 11 HV buses, 272 MV buses and 49 FPU. The size of the MV grids plays the largest role in the computation time, as MV Feeder 4 takes considerably longer to compute than the rest. The computation of the FOR of the HV Grid includes the NR-PF calculations to update the grid voltages.

Table 6-2: Summary of average computation times of process stages and the total processing time (average of 20 repetitions of the experiment).

	Grid	Step 4 / s	Step 7 / s
Iteration 1	MV Feeder 4	4.34	4.16
	MV Feeder 5	1.51	1.51
	MV Feeder 6	1.85	1.83
	HV Grid	1.84	1.85
Total Avg. Processing Time		18.89	

The second benefit is that it gives the possibility to distribute (and parallelize) the computation of the FOR in case the different grid sections are performed by a different grid operator for their own grid sections. This would allow to avoid grid operators from exchanging any additional or sensitive information about the grid topology, other than the voltage and the FOR at the interconnection points.

If the entire grid is controlled by a single entity, parallel FOR computations could lead to a substantial reduction of the computational time, as the total required computation time would depend just on the slowest calculation. Such time reductions can make it possible to apply the method described in Chapter 6.1 for even larger feeders. However, for this thesis, the implementation was performed in a single workstation, thus there are no data transfer delays nor errors, which may slow down the application of the method in real operation. This scenario assumed little information about the initial state of the grid, meaning that a voltage correction stage becomes necessary. In real operation, real-time voltage and current measurements at the MV/HV transformers are usually available nowadays, as well as

---

[18] After 20 repetitions of the experiment

[19] With maximal 128 FOR boundary points and approximating the maximal branch flow with a 128-sides regular polygon.

in MV/LV transformers. A correction stage may not be necessary in every case; however, this depends on the characteristics of each grid.

6.3 Redispatch Concept

The previous use cases show how the FOR concept could be used with time-series and how the bottom-up computation can be distributed over different voltage levels. They allow the visualization of present and future flexibility provision capability of a grid in a methodical way. However, having a graphical representation of the state of the grid is not the only functionality. Some approaches to use the FOR in the control of microgrids were proposed in [79], [113], [148] and [150], or for the economical optimization of a VPP in [152], each with a different objective in sight. However, not much research has been made so far on how to apply the FOR concept to TSO-DSO or even DSO-DSO decision-making schemes.

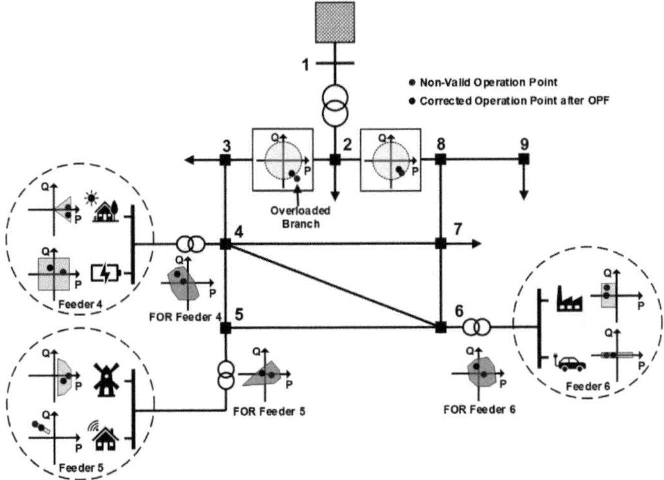

Figure 6-18: Congestion management concept coupled with multi-level aggregation.

In [157], the FOR, simplified as a Type 1 FPU, is used as a constraint of an OPF aiming to reschedule generation for grid balancing purposes. In the given example, the FPG compensate active power production deficit caused by a wind park. Traditionally, in most state-of-the-art OPF formulations, regardless of the goal, the active and reactive power provision of generators is constrained in the form of a Type 1 FPU. Adding the flexibility provision of FPG into the problem is only a matter of interpretation (e.g. [39]), as mathematically, very little is changed, only that the FPG may have a four-quadrant operation, instead of the typical two-quadrant operation of SG, and that the FOR can have any linear convex shape. What was shown in [157], is that the flexibility provided by FPG can be used for grid balancing purposes, however, the bottom-up aggregation process is not completely disclosed. In [55], the FOR concept is used to solve a redispatch problem to resolve a (n-1) violation in the grid caused by a redispatch decision. In this case, the bottom-up computation of the FOR is considered, yet not in a systematic way.

These two publications show how the FOR concept can be incorporated in the operation of transmission and distribution grids, especially as a convex approximated FOR can easily be included into an OPF computation. The issue that remains to be demonstrated, is the viability to reduce the computation time of the process to an extent in which it becomes conceivable to apply it for the everyday operation of the grid. This use case aims to shows how the flexibility from FPG could be used within a short-term redispatch concept. As part of the process, the FOR of each MV feeder (or FPG) is computed and used as input for an OPF at the HV grid, which aims to avoid grid constraint violations. In case of a possible violation, new operation points are computed for each FPG and transmitted back to each FPU. Figure 6-18 shows a schematic representation of the process applied to the MV/HV grids of Chapter 5.1.3, the same one used in this chapter.

6.3.1 Definition of redispatch OPF including FOR constraints

To ensure a stable steady-state operation of the system, an OPF is formulated, which aims at correcting any possible constraint violations by rearranging the operation points of the single FPG within the boundaries given by the FOR. For the sake of demonstration, the objective function of the OPF is a nonlinear multivariable function, while all constraints are linear. Essentially, the OPF defined in Chapter 4.2 for the LFA algorithm is replicated in here, including the linear power flow model and the linear approximations of the branch flow limits. The non-linear objective function seeks to minimize the deviation of the operation points of the FPG compared to their initial operation point for each time-step.

$$min_x f(x) = min_x \sum_{F=1}^{|FPG|} \sqrt{(P_F - P_{F0})^2 + (Q_F - Q_{F0})^2} \qquad (6\text{-}3)$$

$$x = \begin{bmatrix} \vartheta_i \\ u_i \\ P_F \\ Q_F \end{bmatrix}, \forall i \in N, \forall F \in FPG \qquad (6\text{-}4)$$

With: P_F, Q_F Operation point of FPG, active and reactive power

P_{F0}, Q_{F0} Initial operation points of FPG, active and reactive power

$f(x)$ Non-linear objective function

6.3.2 Operating a redispatch concept using FOR

Figure 6-18 shows a scenario in which the grid operation point (red markers) is expected to cause a power line to become overloaded. Having computed the FOR of the underlayered feeders, the proposed OPF is solved and a new set of operation points within the constraints of the FOR are obtained. The new operation point would remedy the situation (blue markers). A snapshot of the grid is shown in the example; however, the process is applied for an entire day (96 periods), following the procedure of Chapter 6.1. And so, the use case merges the two previously described concepts, together with the redispatch concept. Overall, the proposed preventive redispatch is based on the concepts proposed in [55] and [157], whereas the focus is to include the linear FOR constraints in the OPF and to achieve a near real-time operation.

In connection with time-series, the operation points of each grid component, hence of the grid itself, changes over time, meaning that three possible states can happen:

- No grid constraints violation happens; no remedial measures needed.
- FPG can provide enough flexibility and system can be stabilized.
- Algorithm does not converge due to flexibility being not enough or being inadequately placed; requiring additional measures.

6.3.3 Numerical Results

For sake of demonstration, an entire day is simulated (96 periods). In the meshed HV grid of Chapter 5.1.3, all the power that flows through the transformer needs to flow through the power lines connecting buses 2-3 and 2-8. Which for the selected case becomes overloaded during the early morning and midday hours, as shown with the red markers in Figure 6-19, as they are located outside of the quadratic plane defining the maximal branch load of line 2-3.

For each time-step, the three MV feeders are aggregated bottom-up, and the resulting FOR are added to the OPF that is defined at the HV level, described in (6-3) and (6-4). During most parts of the day, the branch is operating normally, requiring no corrective actions. Figure 6-20 shows the operation point changes expected from each FPG in order to solve the congestion in the system.

Because of its placement and significant flexibility potential, MV Feeder 4 needs to provide flexibility more often than the other two feeders. In this scenario, the flexibility provided by the all feeders is enough to correct the congestion. It should be noted, that the HV grid is balanced by the slack bus after the use of flexibility.

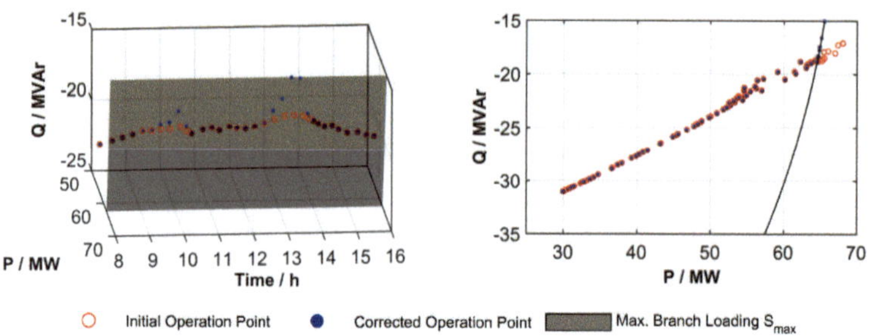

Figure 6-19: Results of congestion management approach for a single day (96 periods).

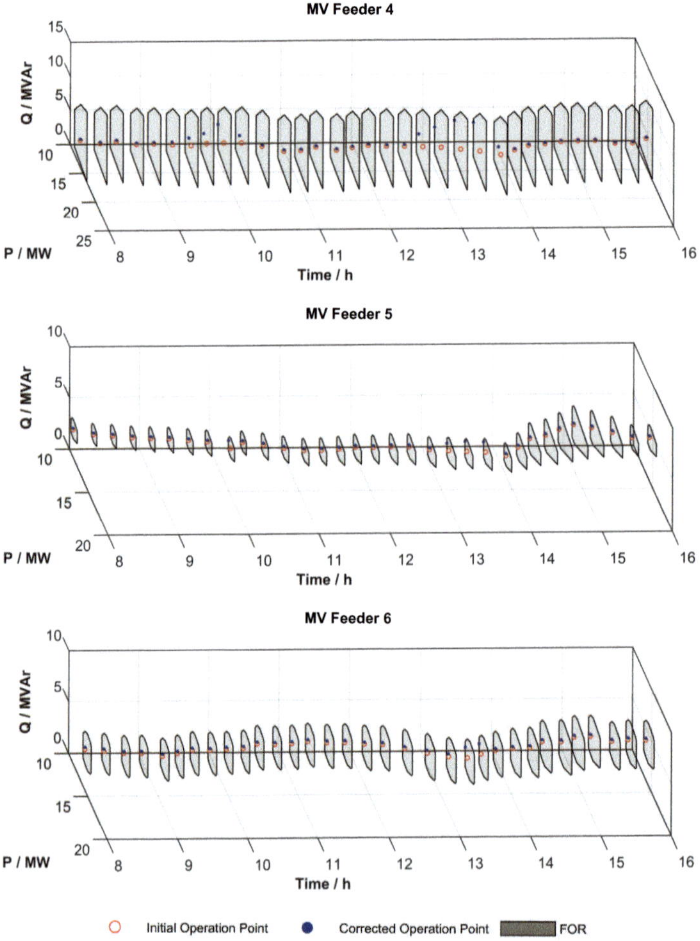

MV Feeder 4

MV Feeder 5

MV Feeder 6

Figure 6-20: Operation point changes within FOR limits of the three MV feeders required to solve the congestion for each time-step.

The objective of the approach is to allow a fast computation of the FOR, permitting a fast preventive congestion management process in general. It was discussed in Chapter 5.7 how the parametrization of the LFA algorithm can affect the computation time. This analysis is complemented here. The process includes the bottom-up aggregation, the voltage correction and the solution of the OPF. An additional step to disaggregate the new operation points of the FPG to the single FPU would still be necessary, however, this is outside the scope of the presented use case. This step requires the definition of an optimization problem with a specific objective function. In a standard load flow computation, one combination of operation points of generators and loads produces a single IPF. However, in the other direction, one IPF can be produced by several combinations of operation points of the single downstream assets. Therefore, additional aspects need to be added to the optimization, in order to obtain a unique set of operation points.

The computation time of the process is evaluated based on two key factors of the LFA algorithm; the number of FOR points that can be evaluated and the number of segments used to linearize the branch flow constraints (same parameter is used for the OPF). Both parameters have an impact on the quality of the provided solution and especially on the computation time, as can be seen in Figure 6-21. Higher values of both parameters would increase the quality of the FOR and the computation time. Generally, 16 or 32 points are enough to describe a FOR, setting a higher bound for the computation time, as it limits the amount of iterations within the LFA algorithm. On the other side, selecting a larger n in Eq. (4-11) would increase the quality of the grid constraint assessment, at the cost of adding more constraints to the optimization problem. This would escalate with the number of buses in the grid, requiring to use a smaller number in the case of larger grids, in order to keep a reduced computation time.

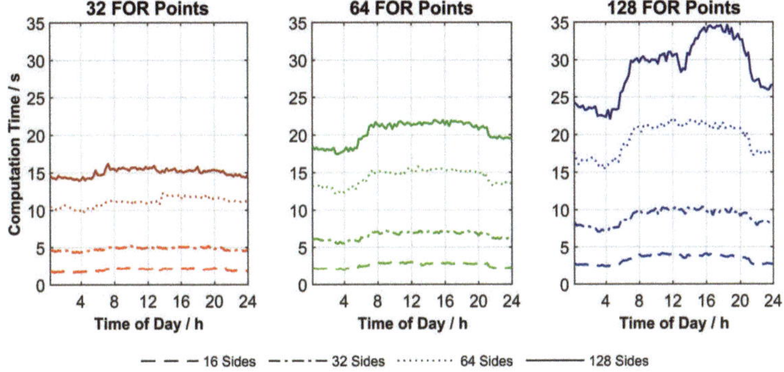

Figure 6-21: Processing time of congestion management approach based on quantity of FOR boundary points and sides of the maximal branch flow approximation.

7 Conclusion and Outlook

7.1 Conclusion

A method to compute the FOR in radial distribution grids was described throughout this thesis, with a focus on the reduction of the computation time. By solving a linear OPF, together with an optimal use of said OPF, it is possible to assess the flexibility provision of realistic MV and LV grids with large number of FPU in short time without reducing the accuracy of the assessed FOR. With the proposed algorithm, and the consequent reduction of the computational time, the integration of the FOR concept in grid operation tasks can become a reality. This offers a new mechanism to improve TSO-DSO coordination, in times when the vertical provision of flexibility is becoming critical for the overall stability of power systems.

The thesis begins in Chapter 2 with a definition of the concept of flexibility. It becomes clear that flexibility has multiple interpretations and representations, aiming at many different use cases. Additionally, there are no universal definitions about the topic; needing first to consolidate existing concepts, as well to defining new ones. This led to the definition of the FPU concept, labelling devices that can offer active and reactive power flexibility as a reaction to an external control signal. It is described how an FPU can be connected to the grid and supply power for ancillary services within the limitations of their capability charts.

The FOR is not a new concept, yet a widely accepted definition was just recently provided in [7]. This definition was adopted in this work and complemented with the FPU idea. In Chapter 3, the FOR is illustrated using an analogy to the classical representation of the capability charts of SG. A handful of methods for the computation of the FOR were analyzed, including analytic, Minkowski sums, random sampling and optimization-based approaches. The general consensus of the discussed methods is that the processing time is still too large to associate the FOR to grid operation tasks.

Based on the analyzed methods, a proposal to compute the FOR in reduced time, i.e. the novel LFA algorithm, was described in Chapter 4. The novelty of the LFA method, compared to other existing methods, is the usage of linear power flow equations instead of the classical nonlinear equations in the OPF, as well as the inclusion of convex linear capability charts of FPU as constraints. A similar approach had been proposed already in [130], however, that model did not consider grid constraints, showcasing the novelty of this thesis. The objective of the LFA method is to be computationally efficient, which is achieved by optimizing and limiting the use of the OPF and by applying parallel computation techniques.

Chapter 5 focusses on the validation of the LFA method, by analyzing its use in different situation and by performing a comparison to other existent techniques (e.g. random sampling and nonlinear OPF). The LFA method is capable of outperforming all analyzed algorithms when applied to grids with a large number of buses and FPU, in both the computation time and the accuracy of the results, as described. These results are complemented with other contributions of the author of this thesis in [4], [108], and [119]. A description of the error added by the linear power flow equations into the assessed FOR was described in Chapter 5.3, yet, the impact in the accuracy is deemed as tolerable in the studied cases. This results in the LFA method showing an excellent trade-off between efficiency and accuracy, fulfilling the main goal of this thesis.

The application of the LFA method to three different use cases focusing on both the planning and the operation of distribution grids was disclosed in Chapter 6. These show how a faster computation of the FOR can help in the development of short- and long-term planning methods considering the vertical flexibility provision of FPU, and in the operation of the grid. The FOR concept can help with the dimensioning of power lines and transformers considering all possible operation states of the grid, and not just the extreme values of the active power flow, as in traditional grid planning. Finally, the integration of the FOR concept in a control strategy for redispatch using flexibility over different voltage levels was demonstrated. These use cases were applied to grid models based on a real power system, therefore, showing its applicability into realistic scenarios.

At the beginning of this work, the following scientific thesis was proposed:

The FOR of distribution grids can be computed as a result of a linear optimization model, decreasing the computation time, while preserving the accuracy; consequently, offering new opportunities to integrate the FOR concept in grid operation processes.

This thesis could be confirmed, with the following outcomes:

- OPF-based FOR computation methods show the best trade-off between efficiency and accuracy among all currently available methods.
- A linear OPF model shows better convergence properties that a non-linear OPF model, allowing for a more accurate and faster calculation of the FOR.
- How the OPF is used during the assessment of the FOR plays an important role in the efficiency of the method.
- Performing multiple FOR computations in reduced time opens new possibilities for its usage in the planning and control of power systems.

7.2   Outlook

Over the course of this dissertation, several aspects of the FOR concept became recurring subjects of discussion at different forums. The first one has to do with the convexity of the linear optimization model. As discussed in 5.4, the "real" FOR can be non-convex under certain conditions. Some precautions need to be taken while applying OPF-based methods to compute the FOR, however, this is strongly dependent on the characteristics of each grid. Nevertheless, as shown in Chapter 5.4 and corroborated in [127] and [128], a convex approximation of the FOR can be forced by applying a linear power flow model with convex constraints. This approximation can in some extreme cases differ greatly from the "real" FOR, however, it ensures that the assessed FOR will still be suitable, as it will only enclose valid IPF. This effect in more complex grid topologies, i.e. meshed grids, heavily loaded grids or grids with large impedances, needs to be further analyzed.

The second conflicting topic relates to the modelling of the flexibility provided by the FPU. The models provided in Chapter 4.3 are intended to provide a more realistic representation of the FOR, compared to widely used simplification of rectangular capability charts. The capability chart of a SG has been exhaustively studied and its overall shape is commonly accepted; however, this is not the case for converter-interfaced generation and storage systems. As described in [43], the technical features of the power inverter play a large role in its capability chart, depending on internal and external factors. This is a critical aspect when considering inverter-based FPU; requiring a more detailed analysis on the quality of the models of the single FPU in order to be able to validate the accuracy of the FOR.

The third aspect relates to the fact that each FOR computation method published so far, including the LFA method, assumes that the grid topology and the characteristics and operation point of the FPU are known. Unfortunately, this is usually not the case in practical applications. At the distribution level, grid measurements are scarce, grid topologies are in some cases not properly documented and there is a vague knowledge of the devices connected behind the meter. Further developments to FOR computation algorithms become necessary before they can be added to grid operation mechanisms. It is crucial to fill the gap regarding grid topology and operation point knowledge, i.e. through state-estimation approaches. One example of a concept following this direction was proposed in [158], in which a state-estimation algorithm is intended to be integrated in the FOR computation process, with a clear focus on real-time operation.

The fourth aspect relates to the use of the FOR as an interface between voltage levels and even among grid operators. A concept for the implementation of a multi-level aggregation approach was described in Chapter 6.2. In reality, there is a tight dependence of the bus voltages over the different grid levels, which in usually neglected in grid modelling with a static slack bus voltage. Changing the operation of a single FPU might cause a more local effect in the grid voltage, however, the FOR computation involves shifting large amounts of power, causing the voltage at the overlayered grid to change. This effect is magnified when several MV grid feeders are added to the calculation, as there is coupling effect between the feeders. This aspect was analyzed in this thesis, however, a more detailed study is necessary, when implementing multi-level approaches in grid with many feeders. A change in the voltage at the overlayered grid can cause the FOR of some underlayered grids to become unfeasible, situation that needs to be properly studied.

The fifth, and final, aspect is related to the usage of the LFA, or similar methods, in the assessment of meshed grids. This question is based on real scenarios, as meshed HV and VHV grids could be operated in parallel with more than one interconnection point between them. This means that the premises of FOR computation methods change drastically, as the power flows do not converge in a single interconnection point (e.g. MV/HV transformer) anymore, instead, they are distributed among all interconnection points (e.g. HV/VHV transformers). This is a current research topic, for which the LFA method cannot offer a direct solution. This issue becomes crucial when it comes to the implementation of the FOR concept in some TSO/DSO schemes.

By the time this thesis was being submitted, two new linear approaches to compute the FOR were published in [106] and [127]. Both papers propose alternative linearization approaches for the power flow equations, while the algorithms follow the same objectives as the LFA method. The provided computational times are comparable to the ones shown in this thesis for the LFA method. This shows the relevance of the LFA method and that linear OPF formulations provide wide-ranging new possibilities to the further development of the FOR concept.

It was clearly stated in [7] that the computational time of the FOR calculation needs to be reduced, otherwise it is difficult to extend its usage to the operation of power systems. With the proposed LFA method, it is demonstrated that it is possible to reduce the computation time of the FOR with an acceptable trade-off with the accuracy. Therefore, the next logical step is to develop use cases where the fast FOR computation can be exploited. Some of which are currently being developed within the *flexQgrid*[20] project, including the integration of state-estimation methods to complement the grid topology with real measurement data.

---

[20] https://flexqgrid.de/

8   References

[1]   J. Cohn, "When the Grid Was the Grid: The History of North America's Brief Coast-to-Coast Interconnected Machine," *Proceedings of the IEEE,* vol. 107, no. 1, pp. 232-243, Jan. 2019.

[2]   S. Howell, Y. Rezgui, J.-L. Hippolyte, B. Jayan and H. Li, "Towards the Next Generation of Smart Grids: Semantic and Holonic Multiagent Management of Distributed Energy Resources," *Renewable and Sustainable Energy Reviews,* vol. 77, pp. 193-214, 2017.

[3]   R. Hidalgo, C. Abbey and G. Joós, "A Review of Active Distribution Networks Enabling Technologies," in *IEEE PES General Meeting,* Providence, RI, USA, Jul. 2010.

[4]   D. A. Contreras and K. Rudion, "Improved Assessment of the Flexibility Range of Distribution Grids Using Linear Optimization," in *Power Systems Computing Conference (PSCC),* Dublin, Ireland, 2018.

[5]   I. Ilic, A. Viskovic and M. Vrazic, "User P-Q Diagram as a Tool in Reactive Power," in *International Conference on the European Energy Market (EEM),* Zagreb, Croatia, May 2011.

[6]   J. R. de Silva, Capability Charts of Power Systems, Canterbury, New Zealand: University of Canterbury, Mar. 1987.

[7]   D. Mayorga, J. Hachenberger, J. Hinker, F. Rewald, U. Häger, C. Rehtanz and J. Myrzik, "Determination of the Time-Dependant Flexibility of Active Distribution Networks to Control Their TSO-DSO Interconnection Power Flow," in *Power Systems Computing Conference (PSCC),* Dublin, Ireland, 2018.

[8]   A. Losi, F. Rossi, M. Russo and P. Verde, "New Tool for Reactive Power Planning," *IEE Proceedings,* vol. 140, no. 4, pp. 256-262, Jul. 1993.

[9]   E. Chiodo, A. Losi, R. Mongelluzzo and F. Rossi, "Capability Charts for Electrical Power Systems," *IEEE Proceedings-c,* vol. 189, no. 1, pp. 71-75, Jan. 1992.

[10]   S. M. Abdelkader and D. Flynn, "Graphical Determination of Network Limits for Wind Power Integration," *IET Generation, Transmission & Distribution,* vol. 3, no. 9, pp. 841-849, 2009.

[11] P. Cuffe, P. Smith and A. Keane, "Capability Chart for Distributed Reactive Power Resources," *IEEE Transactions on Power Systems,* vol. 29, no. 1, pp. 15-22, Jan. 2014.

[12] P. Goergens, F. Portratz, M. Gödde and A. Schnettler, "Determination of the Potential to Provide reactive Power from Distribution Grids to the Transmission Grid Using Optimal Power Flow," in *International Universities Power Engineering Conference (UPEC)*, Soke on Trent, UK, Sep. 2015.

[13] A. V. Jayawardena, L. G. Meegahapola, D. A. Robinson and S. Perera, "Capability Chart: A New Tool for Grid-Tied Microgrid Operation," in *IEEE PES T&D Conference and Exposition*, Chicago, USA, Apr. 2014.

[14] S. Riaz and P. Mancarella, "On Feasibility and Flexibility Operating Regions of Virtual Power Plants and TSO/DSO Interfaces," in *Powertech*, Milan, Italy, Jun. 2019.

[15] EURELECTRIC, "Flexibility and Aggregation Requirements for their interaction in the market," January 2014.

[16] I. Talavera, S. Stepanescu, F. Bennewitz, J. Hanson, R. Huber, F. Oechsle and H. Abele, "Vertical Reactive Power Flexibility based on Different Reactive Power Characteristics for Distributed Energy Resources," in *International ETG Congress*, Bonn, Germany, Nov. 2017.

[17] A. Keane, E. Diskin, P. Cuffe, D. Brooks, T. Hearne and T. Fallon, "Reactive Power Support from Distributed Generation - Ireland's Demonstration Initiative," in *IEEE PES General Meeting*, San Diego, CA, USA, Jul. 2012.

[18] P. Vermeyen and P. Lauwers, "Managing Reactive Power in MV Distribution Grids Containing Distributed Generation," in *CIRED*, Lyon, France, Jun. 2015.

[19] M. Tomaszewski, S. Stankovic, I. Leisse and L. Söder, "Minimization of Reactive Power Exchange at the DSO/TSO Interface: Öland Case," in *ISGT Europe*, Bucharest, Romania, Sep. 2019.

[20] P. Cuffe, P. Smith and A. Keane, "Characterisation of the Reactive Power Capability of Diverse Distributed Generators: Toward an Optimisation Approach," in *IEEE PES General Meeting*, San Diego, CA, USA, Jul. 2012.

[21] M. Heleno, R. Soares, J. Sumaili, R. J. Bessa, L. Seca and M. A. Matos, "Estimation of the Flexibility Range in the Transmission-Distribution Boundary," in *PowerTech*, Eindhoven, Netherlands, 2015.

[22] J. Silva, J. Sumailli, R. Bessa, L. Seca, M. Matos, V. Miranda, M. Cajoulle, B. Goncer and M. Sebastian-Viana, "Estimating the Active and Reactive Power Flexibility Area at the TSO-DSO Interface," *IEEE Transactions on Power Systems,* vol. 33, no. 5, pp. 4741-4750, Sep. 2018.

[23] J. Silva, J. Sumaili, R. J. Bessa, L. Seca, M. Matos and V. Miranda, "The challenges of estimating the impact of distributed energy resources flexibility on the TSO/DSO boundary node operating points," *Computers and Operations Research,* vol. 96, pp. 294-304, Aug. 2018.

[24] N. Fonseca, J. Silva, A. Silva, J. Sumailli, L. Seca, R. Bessa, J. Pereira, M. Matos, P. Matos, A. Morais, M. Caujolle and M. Sebastian-Viana, "EvolvDSO Grid Management Tools to Support TSO-DSO Cooperation," in *CIRED,* Helsinki, Finland, 2016.

[25] M. Braun, PhD Thesis: Provision of Ancillary Services by Distributed Generators, Kassel, Germany: University of Kassel, 2008.

[26] E. Lannoye, D. Flynn and M. O'Malley, "Evaluation of Power System Flexibility," *IEEE Transactions on Power Systems,* vol. 27, no. 2, pp. 922-931, May 2012.

[27] R. Bärenfänger, E. Drayer, D. Daniluk, B. Otto, E. Vanet, R. Caire, T. Shamsi Abbas and B. Lisanti, "Classifying Flexibility Types in Smart Electric Distribution Grids," in *CIRED Workshop,* Helsinki, Finland, 2016.

[28] J. Zhao, T. Zheng and E. Litvinov, "A Unified Framework for Defining and Measuring Flexibility in Power System," *IEEE Transactions on Power Systems,* vol. 31, no. 1, pp. 339-347, Jan. 2016.

[29] A. Ulbig and G. Andersson, "Analyzing Operational Flexibility of Electric Power Systems," *International Journal of Electrical Power & Energy Systems,* vol. 72, pp. 155-164, Nov. 2015.

[30] H. Nosair and F. Bouffard, "Flexibility Envelopes for Power System Operational Planning," *IEEE Transactions on Sustainable Energy,* vol. 6, no. 3, pp. 800-809, Jul. 2015.

[31] North American Electric Reliability Corporation (NERC), "Flexibility Requirements and Potential Metrics for Variable Generation: Implications for System Planning Studies," NERC, Princeton, NJ, USA, Aug. 2010.

[32] K. Heussen, S. Koch, A. Ulbig and G. Andersson, "Energy Storage in Power System Operation: The Power Nodes Modeling Framework," in

IEEE PES Innovative Smart Grid Technologies Conference Europe (ISGT Europe), Gothenberg, Sweden, Oct. 2010.

[33] T. Kornrumpf and M. Zdrallek, "Analyzing Flexibility Options with Cross-Sectoral Requirements on Distribution Level," in *IEEE PES General Meeting*, Portland, OR, USA, Aug. 2018.

[34] A. A. Jahromi and F. Bouffard, "On the Loadability Sets of Power Systems - Part I: Characterization," *IEEE Transactions on Power Systems,* vol. 32, no. 1, pp. 137-145, Jan Jan. 2017.

[35] J. Su and H.-D. Chiang, "Toward Characterization of the Feasible Region of Loadability of Power Systems," in *IEEE PES General Meeting*, Atlanta, GA, USA, Aug. 2019.

[36] M. Bucher, S. Chatzivasileiadis and G. Andersson, "Managing Flexibility in Multi-Area Power Systems," *IEEE Transactions on Power Systems,* vol. 31, no. 2, pp. 1218-1226, Mar. 2016.

[37] M. A. Bucher, S. Delikaraoglou, K. Heussen, P. Pinson and G. Andersson, "On Quantification of Flexibility in Power Systems," in *IEEE PowerTech*, Eindhoven, Netherlands, 2015.

[38] A. Kalantari, J. F. Restrepo and F. D. Galiana, "Security-Constrained Unit Commitment With Uncertain Wind Generation: The Loadability Set Approach," *IEEE Transactions on Power Systems,* vol. 28, no. 2, pp. 1787-1796, May 2013.

[39] G. Petretto, M. Cantú, G. Gigliucci, F. Pilo, G. Pisano, N. Natale, G. G. Soma, M. Coppo and R. Turri, "Representative Distribution Network Models for Assessing the Role of Active Distribution Systems in Bulk Ancillary Services Markets," in *Power Systems Computation Conference (PSCC)*, Genoa, Italy, Aug. 2016.

[40] D. Contreras and K. Rudion, "CALLIA Deliverable Task 1.1 - Part B: Simulations on Use Cases for the Usage of Flexibilities in Distribution Grids," Jul. 2018. [Online]. Available: https://callia.info/wp-content/uploads/2018/09/20180628_D1_1b_v2.pdf. [Accessed 27 02 2020].

[41] D. Contreras, O. Laribi, M. Banka and K. Rudion, "Assessing the Flexibility Provision of Microgrids in MV Distribution Grids," in *CIRED Workshop*, Ljubljana, 2018.

[42] B. M. Buchholz and Z. Styczynski, Smart Grids – Fundamentals and Technologies in Electricity Networks, Heidelberg, Germany: Springer, 2014.

[43] D. Bernet, L. Stefanski and M. Hiller, "A Hybrid Medium Voltage Multilevel Converter with Parallel Voltage-Source Active Filter," in *10th International Conference on Power Electronics - ECCE Asia*, Busan, Korea, May 2019.

[44] D. Pudjianto, C. Ramsay and G. Strbac, "Virtual Power Plant and System Integration of Distributed Energy Resources," *IET Renewable Power Generation,* vol. 1, no. 1, pp. 10-16, 2007.

[45] VDE, "VDE-AR-N 4105 - Power Generating Plants in the Low Voltage Network," VDE, Berlin, Germany, 2019.

[46] VDE, „VDE-AR-N 4110 - Technical Connection Rules for Medium-Voltage," VDE, Berlin, Germany, 2019.

[47] International Energy Agency (IEA), "Do It Locally: Local Voltage Support by Distributed Generation – A Management Summary," IEA, Germany, Jan. 2017.

[48] M. Stötzer, P. Gronstedt, Z. Styczynski, B. M. Buchholz, W. Glaunsinger and K. V. Suslov, "Demand Side Integration - A Potential Analysis for the German Power System," in *IEEE PES General Meeting*, San Diego, CA, USA, Jul. 2012.

[49] R. D'hulst, W. Labeeuw, B. Beusen, S. Claessens, G. Deconinck and K. Vanthournout, "Demand Response Flexibility and Flexibility Potential of Residential Smart Appliances: Experiences from Large Pilot Test in Belgium," *Applied Energy,* vol. 155, pp. 79-90, 2015.

[50] SmartNet, "D1.1 Ancillary Service Provision by RES and DSM Connected at Distribution Level in the Future Power System," Dec. 2012.

[51] dena, "Ancillary Services Study 2030," Berlin, Germany, Jul. 2014.

[52] H. Gerard, E. Rivero and D. Six, "Coordination Between Transmission and Distribution System Operators in the Electricity Sector – A Conceptual Framework," *Utilities Policy,* vol. 50, pp. 40-48, 2019.

[53] E. Rivero, D. Six, A. Ramos and M. Maenhoudt, "Deliverable 1.3 - Preliminary assessment of the future roles of DSOs, future market

architectures and regulatory frameworks for network integration of DRES," 2014. [Online]. Available: http://www.evolvdso.eu.

[54] C. Zhang, Y. Ding, N. C. Nordentoft, P. Pinson and J. Østergaard, "FLECH: A Danish Market Solution for DSO Congestion Management Through DER Flexibility Services," *Journal of Modern Power Systems and Clean Energy,* vol. 2, no. 2, pp. 126-133, 2014.

[55] R. Schwerdfeger, PhD Thesis: Vertikaler Netzbetrieb - Ein Ansatz zur Koordinierung von Netzbetriebinstanzen verschiedener Netzebenen, Ilmenau, Germany: Universitätsverlag Ilmenau, 2017.

[56] D. Westermann, M. Wolter, P. Komarnicki, S. Schlegel, R. Schwerdfeger, A. Richter and B. Arendarski, "Control Strategies for a Fully RES Based Power System," in *International ETG Congress 2017*, Bonn, Germany, Nov. 2017.

[57] O. Pohl, F. Rewald, S. Dalhues, P. Jörke, C. Rehtanz, C. Wietfeld, A. Kubis, R. Kentchim and D. Kirsten, "Advancements in Distributed Power Flow Control," in *2018 53rd International Universities Power Engineering Conference (UPEC)*, Glasgow, UK, Sep. 2018.

[58] A. Hermann, PhD Thesis: Market-Based Methods For The Coordinated Use Of Distributed Energy Resources - TSO-DSO Coordination in Liberalized Power Systems, Kongens Lyngby, Denmark: Technical University of Denmark, 2019.

[59] E. Diskin, P. Cuffe and A. Keane, "Distribution System Reactive Power Management Under Defined Power Transfer Standards," in *IEEE PES General Meeting*, Vancouver, BC, Canada, Jul. 2013.

[60] J. R. de Silva, C. P. Arnold and J. Arrillaga, "Capability Chart for an HVDC Link," *IEE PRoceedings,* vol. 134C, no. 3, pp. 181-186, May 1987.

[61] N. E. Nilsson and J. Mercurio, "Synchronous Generator Capability Curve Testing and Evaluation," *IEEE Transactions on Power Delivery,* vol. 9, no. 1, pp. 414-424, Jan. 1994.

[62] A. Losi, M. Russo, P. Verde and D. Menniti, "Capability Chart for Generator-transformer Units," in *MELECON*, Bari, Italy, Aug. 1996.

[63] G. Gargiulo, V. Mangoni and M. Russo, "Capability Charts for Combined Cycle Power Plants," *IEE Proceedings Generation Transmission Distribution,* vol. 149, no. 4, pp. 407-415, Jul 2002.

[64] A. M. Noman, A. A. Al-Shamma'a, K. E. Addoweesh, A. A. Alabduljabbar and A. I. Alolah, "Simulation and Comparison of Three Phase CHB MLI and Three Phase Cascaded Voltage Source MLI Topologies for Grid Connected PV Applications," in *IECON 2017*, Beijing, China, Nov. 2017.

[65] G. Ertasgin, D. Whaley, N. Ertugrul and W. L. Soong, "A Current-Source Grid-Connected Converter Topology for Photovoltaic Systems," in *AUPEC 2006*, Melbourne, Australia, Dec. 2006.

[66] E. Enrique, "Use of Capability Curves for the Analysis of Reactive Power Compensation in Solar Farms," in *I&CPS 2017*, Niagara Fallas, ON, Canada, May 2017.

[67] R. Albarracín and M. Alonso, "Photovoltaic Reactive Power Limits," in *International Conference on Environment and Electrical Engineering*, Wroclaw, Poland, May 2013.

[68] A. Cabrera-Tobar, E. Bullich-Massagué, M. Aragües-Peñalba and O. Gomis-Bellmunt, "Reactive Power Capability Analysis of a Photovoltaic Generator for Large Scale Power Plants," in *IET RPG 2016*, London, UK, Sep. 2016.

[69] F. Delfino, R. Procopio, M. Rossi and G. Ronda, "Integration of Large-Size Photovoltaic Systems into the Distribution Grids: a P–Q Chart Approach to Assess Reactive Support Capability," *IET Renewable Power Generation*, vol. 4, no. 4, pp. 329-340, 2010.

[70] A. Ellis, R. Nelson, E. Von Engeln, R. Walling, J. MacDowell, L. Casey, E. Seymour, W. Peter, C. Barker, B. Kirby and J. R. Williams, "Reactive Power Performance Requirements for Wind and Solar Plants," in *IEEE PES General Meeting*, San Diego, USA, 2012.

[71] M. Ivas, A. Marušić, J. G. Havelka and I. Kuzle, "P-Q Capability Chart Analysis of Multi-inverter Photovoltaic Power Plant," *Electrical Power and Energy Systems,* vol. 116, 2020.

[72] T. Ackermann, Wind Power in Power Systems, Stockholm, Sweden: John Wiley & Sons, Ltd., 2005.

[73] IEC, *IEC 61400-27-1 Electrical Simulation Models - Wind Turbines,* Geneva, Switzerland: IEC, 2015.

[74] M. E. Montilla-DJesus, S. Arnaltes, E. D. Castronuovo and D. Santos-Martin, "Optimal Power Transmission of Offshore Wind Power Using a VSC-HVDC Interconnection," *Energies,* vol. 10, no. 1046, pp. 1-16, 2017.

[75] A. P. Tennakoon, J. B. Ekanayake, A. Atputharajah and S. G. Abeyratne, "Capability Chart of a Doubly Fed Induction Generation Based on its Ratings and Stability Margin," in *International Universities Power Engineering Conference (UPEC),* Cardiff, UK, Sep. 2010.

[76] J. Tian, C. Su and Z. Chen, "Reactive Power Capability of the Wind Turbine with Doubly Fed Induction Generator," in *IECON 2013*, Vienna, Austria, Nov. 2013.

[77] T. Lund, P. Sørensen and J. Eek, "Reactive Power Capability of a Wind Turbine with Doubly Fed Induction Generator," *Wind Energy,* vol. 10, pp. 379-394, 2007.

[78] M. El Achkar, R. Mbayed, G. Salloum, S. Le Ballois and E. Monmasson, "Generic Study of the Power Capability of a Cascaded Doubly Fed Induction Machine," *Electrical Power and Energy Systems,* vol. 86, pp. 61-70, 2017.

[79] L. Meegahapola, T. Littler and S. Perera, "Capability Curve Based Enhanced Reactive Power Control Strategy for Stability Enhancement and Network Voltage Management," *Electrical Power and Energy Systems,* vol. 52, pp. 96-106, 2013.

[80] I. Erlich, M. Wilch and C. Feltes, "Reactive Power Generation by DFIG Based Wind Farms with AC Grid Connection," in *IEEE Powertech*, Lausanne, Switzerland, 2008.

[81] J. A. Martin and I. A. Hiskens, "Reactive Power Limitation due to Wind-Farm Collector Networks," in *IEEE Powertech*, Eindhoven, Netherlands, Jul. 2015.

[82] M. Sarkar, M. Altin, P. E. Sørensen and A. D. Hansen, "Reactive Power Capability Model of Wind Power Plant Using Aggregated Wind Power Collection System," *Energies,* vol. 12, no. 1607, pp. 1-19, 2019.

[83] N. R. Ullah, K. Bhattacharya and T. Thiringer, "Wind Farms as Reactive Power Ancillary Service Providers—Technical and Economic Issues," *IEEE Transactions on Energy Conversion,* vol. 24, no. 3, pp. 661-672, Sep. 2009.

[84] A. Ríos, S. Arnaltes and J. L. Rodríguez-Amenedo, "Reactive Capability Limits of Wind Farms," *Int. J. Energy Technology and Policy,* vol. 3, no. 3, pp. 213-222, 2005.

[85] E. Enrique, "Generation Capability Curves for Wind Farms," in *SusTech 2014*, Portland, OR, USA, Jul. 2014.

[86] M. H. Hassam, D. Helmi, M. Elshahed and H. Abd-Elkhalek, "Improving the Capability Curves of a Grid-Connected Wind Farm: Gabel El-Zeit, Egypt," in *MEPCON 2017*, Menoufia, Egypt, Dec. 2017.

[87] S. Mondal and D. Kastha, "Maximum Active and Reactive Power Capability of a Matrix Converter-Fed DFIG-Based Wind Energy Conversion System," *IEEE Journal of Emerging and Selected Topics in Power Electronics,* vol. 5, no. 3, pp. 1322-1333, 2017.

[88] P. Cuffe, P. Smith and A. Keane, "Transmission System Impact of Wind Energy Harvesting Networks," *IEEE Transactions on Sustainable Energy,* vol. 3, no. 4, pp. 643-651, Oct. 2012.

[89] S. Engelhardt, I. Erlich, C. Feltes, J. Kretschmann and F. Shewarega, "Reactive Power Capability of Wind Turbines Based on Doubly Fed Induction Generators," *IEEE Transactions on Energy Conversion,* vol. 26, no. 1, pp. 364-372, Mar. 2011.

[90] N. Dinic, B. Fox, D. Flynn, L. Xu and A. Kennedy, "Increasing Wind Farm Capacity," *IEE Proc.-Gener. Transm. Distrib.,* vol. 153, no. 4, pp. 493-498, 2006.

[91] M. Džamarija, M. Bakhtvar and A. Keane, "Operational Characteristics of Non-firm Wind Generation in Distribution Networks," in *IEEE PES General Meeting*, San Diego, CA, USA, Jul. 2012.

[92] O. Krishan and S. Suhag, "An Updated Review of Energy Storage Systems: Classification and Applications in Distributed Generation Power Systems Incorporating Renewable Energy Resources," *International Journal of Energy Research,* vol. 43, pp. 6171-6210, 2019.

[93] D. O. Akinyele and R. K. Rayudu, "Review of Energy Storage Technologies for Sustainable Power Networks," *Sustainable Energy Technologies and Assessments,* vol. 8, pp. 74-91, 2014.

[94] L. Chang, W. Zhang, S. Xu and K. Spence, "Review on Distributed Energy Storage Systems for Utility Applications," *CPSS Transactions on Power Electronics and Applications,* vol. 2, no. 4, pp. 267-276, Dec. 2017.

[95] N. W. Miller, R. S. Zrebiec, R. W. Delmerico and G. Hunt, "Design and Commissioning of a 5MVA, 2.5 MWH Battery Energy Storage System," in *IEEE T&D Conference and Exposition*, Los Angeles, CA, USA, Sep. 1996.

[96] A. Gabash and P. Li, "Evaluation of Reactive Power Capability by Optimal Control of Wind-Vanadium Redox Battery Stations in Electricity Market," in *International Conference on Renewable Energies and Power Quality (ICREPQ)*, Las Palmas de Gran Canaria, Spain, Apr. 2011.

[97] S. Kundu, K. Kalsi and S. Backhaus, "Approximating Flexibility in Distributed Energy Resources: A Geometric Approach," in *Power Systems Computing Conference (PSCC)*, Dublin, Ireland, Jun. 2018.

[98] G. K. Irungu, A. O. Akumu and D. K. Murage, "Modeling Industrial Load Due to Severe Voltage Surges and Sags: 'A Case Study of Magadi Soda Company'," in *AFRICON*, Windhoek, South Africa, 2007.

[99] C. Sourkounis and P. Tourou, "Grid Code Requirements for Wind Power Integration in Europe," in *Power Options for the Eastern Mediterranean Region (POEM)*, Limassol, Cyprus, Nov. 2012.

[100] N. I. Sarkar, L. G. Meegahapola and M. Datta, "Reactive Power Management in Renewable Rich Power Grids: A Review of Grid-Codes, Renewable Generators, Support Devices, Control Strategies and Optimization Algorithms," *IEEE Access,* vol. 6, pp. 41458-41489, May 2018.

[101] A. Ellis, R. Nelson, E. Von Engeln, R. Walling, J. MacDowell, L. Casey, E. Seymour, W. Peter, C. Barker, B. Kirby and J. R. Williams, "Review of Existing Reactive Power Requirements for Variable Generation," in *IEEE PES General Meeting*, San Diego, CA, USA, Jul. 2012.

[102] IEEE, "1547-2018 - IEEE Standard for Interconnection and Interoperability of Distributed Energy Resources with Associated Electric Power Systems Interfaces," IEEE, 2018.

[103] M. Rossi, G. Vigano, D. Moneta, M. T. Vespucci and P. Pisciella, "Fast Estimation of Equivalent Capability for Active Distribution Networks," in *CIRED*, Glasgow, Scotland, Jun. 2017.

[104] M. Sarstedt, L. Kluß, J. Gerster, T. Meldau und L. Hofmann, „Survey and Comparison of Optimization-Based Aggregation Methods for the Determination of the Flexibility Potentials at Vertical System Interconnections," *Energies,* Bd. 14, Nr. 687, Jan. 2021.

[105] A. Losi, F. Rossi, M. Russo and P. Verde, "Capability Chart for the Planning of Reactive Power Resources," in *MELECON*, Bari, Italy, May 1996.

[106] Z. Tan, H. Zhong, Q. Xia, C. Kang, X. Wang and H. Tang, "Estimating the Robust P-Q Capability of a Technical Virtual Power Plant under Uncertainties," *IEEE Transactions on Power Systems,* vol. 35, no. 6, pp. 4285-4296, Nov. 2020.

[107] CAISO, "What the Duck Curve Tells us About Managing a Green Grid," CAISO, Folsom, CA, USA, 2016.

[108] D. A. Contreras and K. Rudion, "Computing the feasible operating region of active distribution networks: Comparison and validation of random sampling and optimal power flow based methods," *IET Generation, Transmission & Distribution,* vol. 15, no. 10, pp. 1600-1612, Jan. May 2021.

[109] E. Handschin, C. Rehtanz, H. F. Wedde, O. Krause and S. Lehnhoff, "On-Line Stable State Determination in Decentralized Power Grid Management," in *Power Systems Computation Conference (PSCC)*, Glasgow, Scotland, Jul. 2008.

[110] O. Krause, S. Lehnhoff, E. Handschin, C. Retanz and H. F. Wedde, "On Feasibility Boundaries of Electrical Power Grids in Steady State," *Electrical Power and Energy Systems,* vol. 31, pp. 437-444, 2009.

[111] R. Schneider, Convex bodies: the Brunn–Minkowski theory, Cambridge, UK: Cambridge University Press, 2013.

[112] S. Das und S. Sarvottamananda, „Computing the Minkowski Sum of Convex Polytopes in Rd," Nov. 2018.

[113] A. Bernstein, J.-Y. Le Boudec, M. Paolone, L. Reyes-Chamorro and W. Saab, "Aggregation of Power Capabilities of Heterogeneous Resources for Real-Time Control of Power Grids," in *Power Systems Computation Conference (PSCC)*, Genoa, Italy, Jun. 2016.

[114] L. Zhao, W. Zhang, H. Hao and K. Kalsi, "A Geometric Approach to Aggregate Flexibility Modeling of Thermostatically Controlled Loads," *IEEE Transactions on Power Systems,* vol. 32, no. 6, pp. 4721-4731, Nov. 2017.

[115] M. S. Nazir, I. A. Hiskens, A. Bernstein and E. Dall'Anese, "Inner Approximation of Minkowski Sums: A Union-Based Approach and Applications to Aggregated Energy Resources," in *IEEE Conference on Decision and Control (CDC)*, Miami Beach, FL, USA, Dec. 2018.

[116] P. Bailly, A. Michiorri and G. Kariniotakis, "Recursive Estimation of Flexibilities in a Radial Distribution Network," in *CIRED*, Madrid, Spain, Jun. 2019.

[117] Z. Yang, H. Zhong, Q. Xia, A. Bose and C. Kang, "Optimal Power Flow Based on Succesive Linear Approximation of Power Flow Equations," *IET Generation Transmission Distribution,* vol. 10, no. 14, pp. 3654-3662, 2016.

[118] L. Ageeva, M. Majidi and D. Pozo, "Analysis of Feasibility Region of Active Distribution Networks," in *2019 International Youth Conference on Radio Electronics, Electrical and Power Engineering (REEPE)*, Moscow, Russia, Mar. 2019.

[119] D. A. Contreras and K. Rudion, "Verification of Linear Flexibility Range Assessment in Distribution Grids," in *Powertech 2019*, Milan, Italy, Jun. 2019.

[120] C. Phillips, J. Anderson, G. Huber and S. Glotzer, "Optimal Filling of Shapes," *Physical Review Letters,* vol. 108, 2012.

[121] H. Blum, A Transformation for Extracting New Descriptors of Shape, MIT Press, 1967, pp. 362-380.

[122] F. Capitanescu, "Critical Review of Recent Advances and Developments Needed in AC Optimal Power Flow," *Electric Power Systems Research,* vol. 136, pp. 57-68, 2016.

[123] A. V. Jayawardena, L. G. Meegahapola, D. A. Robinson and S. Perera, "Microgrid Capability Diagram: A Tool for Optimal Grid-tied Operation," *Renewable Energy,* vol. 74, pp. 497-504, 2015.

[124] H. Chen and A. Moser, "Improved Flexibility of Active Distribution Grid by Remote Control of Renewable Energy Sources," in *International Conference on Clean Electrical Power*, Liguria, Italy, 2017.

[125] F. Capitanescu, "TSO-DSO Interaction: Active Distribution Network Power Chart for TSO Ancillary Services Provision," *Electric Power,* vol. 163, pp. 226-230, Oct. 2018.

[126] J. Buire, F. Colas, J.-Y. Dielout and X. Guillaud, "Stochastic Optimization of PQ Powers at the Interface between Distribution and Transmission Grids," *Energies,* vol. 12, pp. 1-16, Oct. 2019.

[127] M. Kalantar-Neyestanaki, F. Sossan, M. Bozorg and R. Cherkaoui, "Characterizing the Reserve Provision Capability Area of Active Distribution Networks: A Linear Robust Optimization Method," *IEEE Transactions on Smart Grid,* vol. 11, no. 3, pp. 2464-2475, May 2020.

[128] S. Bolognani and S. Zampieri, "On the Existence and Linear Approximation of the Power Flow Solution in Power Distribution Networks," *IEEE Transactions on Power Systems,* vol. 31, no. 1, pp. 163-172, 2016.

[129] G. Valverde and J. J. Orozco, "Reactive Power Limits in Distributed Generators from Generic Capability Curves," in *IEEE PES General Meeting*, National Harbor, MD, USA, 2014.

[130] S. Bolognani und F. Dörfler, „Fast Power System Analysis via Implicit Linearization of the Power Flow Manifold," in *Fifty-third Annual Allerton Conference*, Allerton House, UIUC, Illinois, USA, Sep. 2015.

[131] D. Lee, H. Nguyen, K. Dvijotham and K. Turitsyn, "Convex Restriction of Power Flow Feasibility Sets," *IEEE Transactions on Control of Network Systems,* vol. 6, no. 3, pp. 1235-1245, Sep. 2019.

[132] Z. Yang, H. Zhong, A. Bose, T. Zheng, Q. Xia and C. Kang, "A Linearized OPF Model With Reactive Power and Voltage Magnitude: A Pathway to Improve the MW-Only DC OPF," *IEEE Transactions on Power Systems,* vol. 33, no. 2, pp. 1734-1745, Mar. 2018.

[133] D. Wang, K. Turitsyn and M. Chertkov, "DistFlow ODE: Modeling, Analyzing and Controlling Long Distribution Feeder," in *51st IEEE Conference on Decision and Control*, Maui, HI, USA, Dec. 2012.

[134] Verband der Elektrotechnik Elektronik Informationstechnik e.V. (VDE), "Power Generation Systems Connected to the Low-voltage Distribution," VDE-AR-N 4105:2011-08, Aug. 2011.

[135] H. Kim and W. Kim, "Integrated Optimization of Combined Generation and Transmission Expansion Planning Considering Bus Voltage Limits," *Journal of Electrical Engineering and Technology,* vol. 9, no. 4, pp. 1202-1209, Jul. 2014.

[136] P. Pareek and A. Verma, "Linear OPF with Linearization of Quadratic Branch Flow Limits," in *2018 IEEMA Engineer Infinite Conference (eTechNxT)*, New Delhi, India, Mar. 2018.

[137] K. Volk, L. Rupp, K. Geschermann and C. Lakenbrink, "Managing Local Flexible Generation and Consumption Units Using a Quota-Based Grid Traffic Light Approach," in *CIRED*, Madrid, Spain, Jun. 2019.

[138] R. de Groot, J. Morren and J. Slootweg, "Investigation of Grid Loss Reduction und Closed-Ring Operation of MV Distribution Grids," in *IEEE PES General Meeting*, Ann Harbor, MD, USA, Jul. 2014.

[139] J. McDonald, B. Wojszczyk, B. Flynn and I. Voloh, "Distribution Systems, Substations, and Integration of Distributed Generation," in *Electrical Transmission Systems and Smart Grids*, New York, NY, USA, Springer, 2013, pp. 7-68.

[140] D. A. Contreras and K. Rudion, "Impact of Grid Topology and Tap Position Changes on the Flexibility Provision from Distribution Grids," in *ISGT Europe*, Bucharest, Romania, Oct. 2019.

[141] D. A. Contreras and K. Rudion, "Time-Based Aggregation of Flexibility at the TSO-DSO Interconnection Point," in *IEEE PES General Meeting*, Atlanta, GA, USA, Aug. 2019.

[142] E. Polymeneas and S. Meliopoulos, "Aggregate Modeling of Distribution Systems for Multi-Period OPF," in *Power Systems Computation Conference (PSCC)*, Genoa, Italy, Jun. 2016.

[143] K. Rudion, A. Orths, Z. A. Styczynski and K. Strunz, "Design of Benchmark of Medium Voltage Distribution Network for Investigation of DG Integration," in *IEEE Power Engineering Society General Meeting*, Montreal, Que., Canada, Jun. 2006.

[144] CIGRE, "Benchmark Systems for Network Integration of Renewable and Distributed Energy Resources," CIGRE Task force C6.04.02, 2011.

[145] M. Brunner, PhD Thesis: Auswirkungen von Power-to-Heat in elektrischen Verteilnetzen, Stuttgart, Germany: Universität Stuttgart, 2017.

[146] M. Kraiczy, G. Lammert, T. Stetz, S. Gehler, G. Arnold, M. Braun, S. Schmidt, H. Homeyer, U. Zickler, F. Sommerwerk and C. Elbs, "Parameterization of Reactive Power Characteristics for Distributed

Generators: Field Experience and Recommendations," in *ETG Congress 2015*, Bonn, Germany, Nov. 2015.

[147] B. Gorgan, S. Busoi, G. Tanasescu and P. V. Nothinger, "PV Plant Modeling for Power System Integration Using PSCAD Software," in *International Symposium on Advanced Topics in Electrical Engineering*, Bucharest, Romania, May 2015.

[148] M. Legry, J.-Y. Dieulot, F. Colas, C. Saudemont and O. Ducarne, "Non-linear Primary Control Mapping for Droop-like Behavior of Microgrid Systems," *IEEE Transactions on Smart Grid,* 2020.

[149] R. Schwerdfeger, S. Schlegel, D. Werstermann, M. Lutter and M. Junghans, "Methodology for Next Generation System Operation Between DSO and DSO," in *CIGRE Session*, Paris, France, 2016.

[150] M. Banka, D. Contreras and K. Rudion, "Multi-Agent Based Strategy for Controlled Islanding and System Restoration Employing Dispersed Generation," in *CIRED Workshop*, Berlin, Germany, 2020.

[151] R. Schwerdfeger, S. Schlegel and D. Westermann, "Approach for N-1 Secure Grid Operation with 100% Renewables," in *IEEE PES General Meeting*, Boston, USA, Jul. 2016.

[152] H. Wang, S. Riaz and P. Mancarella, "Integrated Techno-Economic Modeling, Flexibility Analysis, and Business Case," *Applied Energy,* vol. 259, no. 1, 2020.

[153] P. Wiest, D. Gross, K. Rudion and A. Probst, "Rapid Identification of Worst-Case Conditions: Improved Planning of Active Distribution Grids," *IET Generation, Transmission & Distribution,* vol. 11, no. 9, pp. 2412-2417, 2017.

[154] P. Wiest, D. Contreras, D. Groß and K. Rudion, "Synthetic Load Profiles of Various Customer Types for Smart Grid Simulations," in *NEIS Conference*, Hamburg, Germany, 2018.

[155] J. Merino, J. E. Rodriguez-Seco, C. Caerts, K. Visscher, R. D'hulst, E. Rikos and A. Temiz, "Scenarios and Requirements for the Operation of the 2030+ Electricity Network," in *CIRED*, Lyon, France, Jun. 2015.

[156] F. Marten, L. Löwer, J. C. Töbermann and M. Braun, "Optimizing the Reactive Power Balance Between a Distribution and Transmission Grid

Through Iteratively Updated Grid Equivalents," in *Power Systems Computing Conference (PSCC)*, Wroclaw, Poland, Feb. 2015.

[157] F. Capitanescu, "OPF Integrating Distribution Systems Flexibility for TSO Real-Time Active Power Balance Management," in *MEDPOWER 2018*, Cavtat, Croatia, Nov. 2018.

[158] F. Rewald, O. Pohl and C. Rehtanz, "State Estimation and Determination of Flexibility Potential in Medium Voltage Networks," in *International Workshop on Flexibility and Resiliency Problems of Electric Power Systems (FREPS 2019)*, Irkutsk, Russia, Aug. 2019.

[159] R. D. Zimmerman, C. E. Murrillo-Sanchez and R. J. Thomas, "MATPOWER: Steady-State Operations, Planning and Analysis Tools for Power Systems Research and Education," *IEEE Transactions on Power Systems,* vol. 26, no. 1, pp. 12-19, Feb. 2011.

Appendix

A List of Own Publications

2021	-	S. Müller, D. Contreras, H. Früh, and K. Rudion, „Online Aggregation of the Flexibility Potential in Distribution Grids using State Estimation," in *CIRED*, Geneva, Switzerland, Jun. 2021.
	-	D. A. Contreras, S. Müller, and K. Rudion, „Congestion Management Using Aggregated Flexibility at the TSO-DSO Interface," in *IEEE Powertech*, Madrid, Spain, Jul. 2021.
	-	D. A. Contreras and K. Rudion, "Computing the feasible operating region of active distribution networks: Comparison and validation of random sampling and optimal power flow based methods," *IET Generation, Transmission & Distribution*, Jan 2021.
2020	-	K. Walz, D. Contreras, K. Rudion, "Synthetic Charging Profiles of Battery-Electric Trucks for Probabilistic Grid Planning," in *CIRED Workshop*, Berlin; Germany, Jun. 2020.
	-	M. Banka, D. Contreras, and K. Rudion, "Multi-Agent Based Strategy for Controlled Islanding and System Restoration Employing Dispersed Generation," in *CIRED Workshop*, Berlin; Germany, Jun. 2020.
	-	K. Walz, D. Contreras, K. Rudion, and P. Wiest, "Modelling of Workplace Electric Vehicle Charging Profiles based on Trip Chain Generation," in *IEEE ISGT Europe*, Delft, Netherlands, Oct. 2020.
2019	-	D. A. Contreras and K. Rudion, "Time-Based Aggregation of Flexibility at the TSO-DSO Interconnection Point," in *IEEE PES General Meeting*, Atlanta, GA, USA, Aug. 2019.
	-	D. A. Contreras and K. Rudion, "Verification of Linear Flexibility Range Assessment in Distribution Grids," in *IEEE Powertech*, Milan, Italy, Jun. 2019.
	-	D. A. Contreras and K. Rudion, "Analysis of the Influence of Grid Topology Changes on the Flexibility Provision of Distribution Grids," in *IEEE ISGT Europe*, Bucharest, Romania, Sep. 2019.
	-	D. Groß, H. Früh, P. Wiest, D. Contreras, K. Rudion, L. Rupp, and C. Lakenbrink, „Evaluation of a Three-Phase Distribution System State Estimation for Operational Use in a Real Medium Voltage Grid," in *IEEE ISGT Europe 2019*, Bucharest, Romania, Sep. 2019.
2018	-	M. Banka, D. Contreras, and K. Rudion, "Hardware-in-the-loop Test Bench for Investigation of DER Integration Strategies within a multi-agent-based Environment," in *IEEE International Energy Conference (ENERGYCON)*, Limassol, Cyprus, Paper No. 137, Jun. 2018.
	-	D. Contreras, O. Laribi, M. Banka, and K. Rudion, "Assessing the Flexibility Provision of Microgrids in MV Distribution Grids," in *CIRED Workshop*, Ljubljana, Slovenia, paper No. 428, Jun. 2018.

- D. A. Contreras and K. Rudion, "Improved Assessment of the Flexibility Range of Distribution Grids Using Linear Optimization," in *Power Systems Computing Conference (PSCC)*, Dublin, Ireland, Jun. 2018.

- P. Wiest, D. Contreras, D. Groß, and K. Rudion, "Synthetic Load Profiles of Various Customer Types for Smart Grid Simulations," in *Conference on Sustainable Energy Supply and Energy Storage Systems (NEIS)*, Hamburg, Germany, Poster-Session 2, Sep. 2018.

- D. Groß, D. Contreras, H. Früh, and K. Rudion, "Systemischer Einsatz von Flexibilitäten zur optimierten Betriebsführung in Verteilnetzen," in *Ingenieurspiegel*, Nov. 2018.

B Analytical Analysis of Two-bus System Case

This chapter demonstrates equations (3-2) and (3-3), based on [10]. The power line is assumed to be lossless, otherwise the equation system would not be valid, as it assumes that the complex power is the same at both ends of the power line.

$$\underline{E} = \underline{I} \cdot \underline{Z} + \underline{U} = \frac{\underline{S}^*}{\underline{E}^*} \cdot \underline{Z} + \underline{U} \tag{0-1}$$

$$\Rightarrow E = \frac{P - jQ}{E} \cdot Z\angle\theta + U\angle\vartheta \tag{0-2}$$

$$\Rightarrow P - jQ = \frac{(E - U\angle\vartheta) \cdot E}{Z\angle\theta} = \frac{\left(E - U \cdot (\cos\vartheta + j\sin\vartheta)\right) \cdot E}{Z \cdot (\cos\theta + j\sin\theta)} \tag{0-3}$$

$$\Rightarrow P - jQ = \frac{\left(E^2 - E \cdot U \cdot (\cos\vartheta + j\sin\vartheta)\right)}{Z \cdot (\cos\theta + j\sin\theta)} \cdot \frac{(\cos\theta - j\sin\theta)}{(\cos\theta - j\sin\theta)} \tag{0-4}$$

$$\Rightarrow P - jQ = \frac{E^2}{Z} \cdot (\cos\theta - j\sin\theta)$$
$$- \frac{E \cdot U}{Z} \cdot (\cos\vartheta + j\sin\vartheta)(\cos\theta - j\sin\theta) \tag{0-5}$$

$$\Rightarrow P - jQ = \frac{E^2}{Z} \cdot (\cos\theta - j\sin\theta)$$
$$- \frac{E \cdot U}{Z} \cdot (\cos\vartheta\cos\theta - j\cos\vartheta\sin\theta + j\sin\vartheta\cos\theta + \sin\vartheta\sin\theta) \tag{0-6}$$

$$\Rightarrow P - jQ = \frac{E^2}{Z} \cdot (\cos\theta - j\sin\theta) - \frac{E \cdot U}{Z} \cdot (\cos(\theta - \vartheta) - j\sin(\theta - \vartheta)) \tag{0-7}$$

$$\Rightarrow P = \frac{E^2}{Z}\cos\theta - \frac{EU}{Z}\cos(\theta - \vartheta) \tag{0-8}$$

$$\Rightarrow Q = \frac{E^2}{Z}\sin\theta - \frac{EU}{Z}\sin(\theta - \vartheta) \tag{0-9}$$

$$\Rightarrow U \cdot \cos(\theta - \vartheta) = E \cdot \cos(\theta) - \frac{P_G \cdot Z}{E} \tag{0-10}$$

$$\Rightarrow U \cdot \sin(\theta - \vartheta) = E \cdot \sin(\theta) - \frac{Q_G \cdot Z}{E} \tag{0-11}$$

With: $\underline{E} = E\angle 0°$ Complex voltage at slack bus (with reference angle)

         $\underline{I}$ Power line complex current

         $\underline{Z} = Z\angle\theta$ Power line complex impedance

         $\underline{U} = U\angle\vartheta$ Complex bus voltage

         $\underline{S} = P + jQ$ Complex power injected by generator

C Parametrization of Grid Models

Table B-1:   *FPU parametrization of LV Feeder 1*

Bus	Type	P_A	P_B	P_C	P_D	Q_A	Q_B	Q_C	Q_D	Q_E	Q_F
7	3	0.030	0.021	-0.021	-0.030	0.030	0.024	0.018	-0.018	-0.024	-0.030
11	3	0.021	0.015	-0.015	-0.021	0.021	0.017	0.013	-0.013	-0.017	-0.021
16	4	0.006	0.001	0.001	0.000	0.002	-0.002	0.000	0.000	0.000	0.000
17	5	0.004	0.000	0.000	0.000	0.001	-0.001	0.000	0.000	0.000	0.000
19	5	0.003	0.000	0.000	0.000	0.001	-0.001	0.000	0.000	0.000	0.000

Table B-2:   *FPU parametrization of LV Feeder 2*

Bus	Type	P_A	P_B	P_C	P_D	Q_A	Q_B	Q_C	Q_D	Q_E	Q_F
3	5	0.060	0.000	0.000	0.000	0.002	-0.002	0.000	0.000	0.000	0.000
3	3	0.034	0.028	-0.028	-0.034	0.034	0.028	0.021	-0.021	-0.028	-0.034

Table B-3:   *FPU parametrization of LV Feeder 3*

Bus	Type	P_A	P_B	P_C	P_D	Q_A	Q_B	Q_C	Q_D	Q_E	Q_F
4	5	0.040	0.000	0.000	0.000	0.013	-0.013	0.000	0.000	0.000	0.000
5	5	0.035	0.000	0.000	0.000	0.012	-0.012	0.000	0.000	0.000	0.000
8	3	0.030	0.021	-0.021	-0.030	0.030	0.024	0.018	-0.018	-0.024	-0.030
20	5	0.003	0.000	0.000	0.000	0.000	0.000	0.000	0.000	0.000	0.000
21	5	0.015	0.000	0.000	0.000	0.005	-0.005	0.000	0.000	0.000	0.000
21	3	0.004	0.003	-0.003	-0.004	0.004	0.004	0.003	-0.003	-0.004	-0.004

Table B-4:   *FPU parametrization of MV Feeder 4*

Bus	Type	P_A	P_B	P_C	P_D	Q_A	Q_B	Q_C	Q_D	Q_E	Q_F
2	5	0.045	0.009	0.000	0.000	0.015	-0.015	0.003	-0.003	0.000	0.000
6	5	0.006	0.001	0.000	0.000	0.002	-0.002	0.000	0.000	0.000	0.000
9	5	0.040	0.008	0.000	0.000	0.013	-0.013	0.003	-0.003	0.000	0.000
11	5	0.014	0.003	0.000	0.000	0.004	-0.004	0.001	-0.001	0.000	0.000
31	5	0.030	0.006	0.000	0.000	0.010	-0.010	0.002	-0.002	0.000	0.000
34	5	0.037	0.007	0.000	0.000	0.012	-0.012	0.002	-0.002	0.000	0.000
34	5	0.067	0.013	0.000	0.000	0.022	-0.022	0.004	-0.004	0.000	0.000
45	5	0.006	0.001	0.000	0.000	0.002	-0.002	0.000	0.000	0.000	0.000
53	5	0.006	0.001	0.000	0.000	0.002	-0.002	0.000	0.000	0.000	0.000
53	5	0.006	0.001	0.000	0.000	0.002	-0.002	0.000	0.000	0.000	0.000
56	5	0.044	0.009	0.000	0.000	0.015	-0.015	0.003	-0.003	0.000	0.000
57	6	0.414	0.230	0.230	0.046	0.414	0.307	0.201	-0.201	-0.398	-0.460
60	5	0.007	0.001	0.000	0.000	0.002	-0.002	0.000	0.000	0.000	0.000
60	5	0.025	0.005	0.000	0.000	0.008	-0.008	0.002	-0.002	0.000	0.000
62	5	0.019	0.004	0.000	0.000	0.006	-0.006	0.001	-0.001	0.000	0.000
64	6	0.360	0.200	0.200	0.040	0.360	0.267	0.174	-0.174	-0.346	-0.400
66	5	0.065	0.013	0.000	0.000	0.021	-0.021	0.004	-0.004	0.000	0.000
75	6	0.090	0.050	0.050	0.010	0.090	0.067	0.044	-0.044	-0.087	-0.100
78	5	0.045	0.009	0.000	0.000	0.015	-0.015	0.003	-0.003	0.000	0.000
82	6	0.473	0.263	0.263	0.053	0.473	0.351	0.229	-0.229	-0.455	-0.525
85	5	0.105	0.021	0.000	0.000	0.035	-0.035	0.007	-0.007	0.000	0.000
85	5	0.049	0.010	0.000	0.000	0.016	-0.016	0.003	-0.003	0.000	0.000
91	5	0.048	0.010	0.000	0.000	0.016	-0.016	0.003	-0.003	0.000	0.000
103	5	0.690	0.138	0.000	0.000	0.227	-0.227	0.045	-0.045	0.000	0.000
104	6	2.700	1.500	1.500	0.300	2.700	2.004	1.308	-1.308	-2.598	-3.000
107	5	0.044	0.009	0.000	0.000	0.014	-0.014	0.003	-0.003	0.000	0.000
107	6	0.675	0.375	0.375	0.075	0.675	0.501	0.327	-0.327	-0.650	-0.750
108	6	0.675	0.375	0.375	0.075	0.675	0.501	0.327	-0.327	-0.650	-0.750
111	5	0.021	0.004	0.000	0.000	0.007	-0.007	0.001	-0.001	0.000	0.000

Table B-5:   *FPU parametrization of MV Feeder 5*

Bus	Type	P_A	P_B	P_C	P_D	Q_A	Q_B	Q_C	Q_D	Q_E	Q_F
2	5	0.313	0.063	0.000	0.000	0.103	-0.103	0.021	-0.021	0.000	0.000
20	5	0.690	0.138	0.000	0.000	0.227	-0.227	0.045	-0.045	0.000	0.000
23	5	0.010	0.002	0.000	0.000	0.003	-0.003	0.001	-0.001	0.000	0.000
27	5	0.565	0.113	0.000	0.000	0.186	-0.186	0.037	-0.037	0.000	0.000
38	5	0.021	0.004	0.000	0.000	0.007	-0.007	0.001	-0.001	0.000	0.000
39	6	0.641	0.356	0.356	0.000	0.641	0.476	0.310	-0.310	-0.617	-0.712
46	5	0.690	0.138	0.000	0.000	0.227	-0.227	0.045	-0.045	0.000	0.000
73	5	0.690	0.138	0.000	0.000	0.227	-0.227	0.045	-0.045	0.000	0.000
74	5	3.920	0.784	0.000	0.000	1.288	-1.288	0.258	-0.258	0.000	0.000

Table B-6: *FPU parametrization of MV Feeder 6*

Bus	Type	P_A	P_B	P_C	P_D	Q_A	Q_B	Q_C	Q_D	Q_E	Q_F
4	6	0.500	0.300	0.300	0.000	0.500	0.400	0.150	-0.150	-0.400	-0.500
9	5	1.388	0.000	0.000	0.000	0.456	-0.456	0.000	0.000	0.000	0.000
21	5	0.177	0.000	0.000	0.000	0.058	-0.058	0.000	0.000	0.000	0.000
23	5	0.144	0.000	0.000	0.000	0.047	-0.047	0.000	0.000	0.000	0.000
30	5	0.066	0.000	0.000	0.000	0.022	-0.022	0.000	0.000	0.000	0.000
32	5	0.053	0.000	0.000	0.000	0.017	-0.017	0.000	0.000	0.000	0.000
59	5	1.058	0.000	0.000	0.000	0.348	-0.348	0.000	0.000	0.000	0.000
61	5	0.076	0.000	0.000	0.000	0.025	-0.025	0.000	0.000	0.000	0.000
62	5	0.115	0.000	0.000	0.000	0.038	-0.038	0.000	0.000	0.000	0.000
63	5	0.318	0.000	0.000	0.000	0.105	-0.105	0.000	0.000	0.000	0.000
73	6	0.675	0.338	0.338	0.000	0.675	0.501	0.327	-0.327	-0.670	-0.750

D Jacobi Matrix Definition

The Jacobian J of the power flow equations contains the partial derivatives of equations (3-15) and (3-16) with respect to the voltage magnitude and angle $(u_i, \vartheta_i, \forall i \in \{B - slack\})$. This is described as follows:

$$J = \begin{bmatrix} \dfrac{\partial p_i}{\partial \vartheta} & \dfrac{\partial p_i}{\partial u} \\ \dfrac{\partial q_i}{\partial \vartheta} & \dfrac{\partial q_i}{\partial u} \end{bmatrix} = \begin{bmatrix} J_1 & J_2 \\ J_3 & J_4 \end{bmatrix} \tag{0-12}$$

$$J_1 = \begin{cases} \dfrac{\partial p_i}{\partial \vartheta_j} = u_i \cdot u_j \cdot y_{ij} \cdot sin(\vartheta_{ij} - \theta_{ij}), j \neq i \\[2em] \dfrac{\partial p_i}{\partial \vartheta_i} = -u_i \cdot \displaystyle\sum_{\substack{j=1 \\ j \neq i}}^{n} u_j \cdot y_{ij} \cdot sin(\vartheta_{ij} - \theta_{ij}) \end{cases} \tag{0-13}$$

$$J_2 = \begin{cases} \dfrac{\partial P_i}{\partial u_j} = u_i \cdot y_{ij} \cdot cos(\vartheta_{ij} - \theta_{ij}), j \neq i \\[2em] \dfrac{\partial P_i}{\partial u_i} = 2 \cdot u_i \cdot y_{ii} \cdot cos(\theta_{ij}) + \displaystyle\sum_{\substack{j=1 \\ j \neq i}}^{n} u_j \cdot y_{ij} \cdot cos(\vartheta_{ij} - \theta_{ij}) \end{cases} \tag{0-14}$$

$$J_3 = \begin{cases} \dfrac{\partial q_i}{\partial \vartheta_j} = -u_i \cdot u_j \cdot y_{ij} \cdot cos(\vartheta_{ij} - \theta_{ij}), j \neq i \\[2em] \dfrac{\partial q_i}{\partial \vartheta_i} = u_i \cdot \displaystyle\sum_{\substack{j=1 \\ j \neq i}}^{n} u_j \cdot y_{ij} \cdot cos(\vartheta_{ij} - \theta_{ij}) \end{cases} \tag{0-15}$$

$$J_4 = \begin{cases} \dfrac{\partial q_i}{\partial u_j} = u_i \cdot y_{ij} \cdot sin(\vartheta_{ij} - \theta_{ij}), j \neq i \\[2em] \dfrac{\partial q_i}{\partial u_i} = 2 \cdot u_i \cdot y_{ii} \cdot sin(-\theta_{ij}) + \displaystyle\sum_{\substack{j=1 \\ j \neq i}}^{n} u_j \cdot y_{ij} \cdot sin(\vartheta_{ij} - \theta_{ij}) \end{cases} \tag{0-16}$$

The Jacobian of the branch flow equations (3-17), (3-19) and (3-20) with respect to the voltage magnitude and angle (u_i, ϑ_i, $\forall i \in \{B - slack\}$) is describes as:

$$\begin{bmatrix} \dfrac{\partial p_{ij}}{\partial \vartheta} & \dfrac{\partial p_{ij}}{\partial u} \\[2mm] \dfrac{\partial q_{ij}}{\partial \vartheta} & \dfrac{\partial q_{ij}}{\partial u} \end{bmatrix} = \begin{bmatrix} J_5 & J_6 \\ J_7 & J_8 \end{bmatrix} \tag{0-17}$$

$$J_5 = \begin{cases} \dfrac{\partial p_{ij}}{\partial \vartheta_i} = u_i \cdot u_j \cdot y_{ij} \cdot sin(\vartheta_{ij} - \theta_{ij}) \\[3mm] \dfrac{\partial p_{ij}}{\partial \vartheta_j} = -u_i \cdot u_j \cdot y_{ij} \cdot sin(\vartheta_{ij} - \theta_{ij}) \\[3mm] \dfrac{\partial p_{ij}}{\partial \vartheta_k} = 0, \forall k \neq i, j \end{cases} \tag{0-18}$$

$$J_6 = \begin{cases} \dfrac{\partial p_{ij}}{\partial u_i} = 2 \cdot u_i \cdot y_{ij} \cdot cos(\theta_{ij}) - u_j \cdot y_{ij} \cdot cos(\vartheta_{ij} - \theta_{ij}) \\[3mm] \dfrac{\partial p_{ij}}{\partial u_j} = -u_i \cdot y_{ij} \cdot cos(\vartheta_{ij} - \theta_{ij}) \\[3mm] \dfrac{\partial p_{ij}}{\partial u_k} = 0, \forall k \neq i, j \end{cases} \tag{0-19}$$

$$J_7 = \begin{cases} \dfrac{\partial q_{ij}}{\partial \vartheta_i} = -u_i \cdot u_j \cdot y_{ij} \cdot cos(\vartheta_{ij} - \theta_{ij}) \\[3mm] \dfrac{\partial q_{ij}}{\partial \vartheta_j} = u_i \cdot u_j \cdot y_{ij} \cdot cos(\vartheta_{ij} - \theta_{ij}) \\[3mm] \dfrac{\partial q_{ij}}{\partial \vartheta_k} = 0, \forall k \neq i, j \end{cases} \tag{0-20}$$

$$J_8 = \begin{cases} \dfrac{\partial q_{ij}}{\partial u_i} = 2 \cdot u_i \cdot y_{ij} \cdot sin(-\theta_{ij}) - u_j \cdot y_{ij} \cdot sin(\vartheta_{ij} - \theta_{ij}) \\[3mm] \dfrac{\partial q_{ij}}{\partial u_j} = -u_i \cdot y_{ij} \cdot sin(\vartheta_{ij} - \theta_{ij}) \\[3mm] \dfrac{\partial q_{ij}}{\partial u_j} = 0, \forall k \neq i, j \end{cases} \tag{0-21}$$